Also by James Jauncey

The Albatross Conspiracy
(for younger readers)

THE MAPMAKER

by

James Jauncey

André Deutsch

For Sarah – my compass

First published in Great Britain in 1994 by
André Deutsch Limited
106 Great Russell Street, London WC1B 3LJ

Cataloguing-in-publication data
for this title is available from
the British Library

ISBN 0 233 98852 1

Printed in Great Britain by
WBC Bridgend

PART ONE

One

was approaching my twenty-first year when I made my discovery. It was the day after my brother drowned.

We were down by the estuary, my brother and I, cutting reeds at the edge of the lagoon which sweeps back into the marsh. It was mid-winter, dawn, crackling cold. As we stepped onto the ice the geese clouded from their roosting pool and streamed inland towards the rising sun, arranging themselves into honking skeins as they went. I never knew where they were heading although I imagined it was to some place beyond the skyline, beyond my experience, from which they returned every evening, untroubled by what they had seen. Our horizons were small in those days.

My brother ran ahead with all the exuberance of his lesser years, slithering and whooping as he gathered speed. Then he dropped to one knee, flung out his arms like an acrobat and sailed across the ice, sketching twin trails through the fine, white dusting of frost. Revealed, the ice was as smooth as marble and so clear it was black with the water underneath.

He came to a halt some distance away and stood up, a hank of hair across his forehead, his cheeks red as meat in the bitter cold. He beckoned to me and shouted and a puff of steam came from his mouth. Older and more cautious, I advanced hesitantly until confident that my feet would not desert me. Then I also ran, dropped and slid headlong with the wind of my own speed burning my ears.

I ended up near the roosting pool. My brother walked over and punched me on the arm, grinning. He made some joke, some boyish expression of glee which I can no longer recall, but I do remember the wild, infectious exhilaration in his eyes. I grabbed his hat, a moth-eaten thing made of squirrel-skins, and threw it into the pale blue air, blinking as it cartwheeled across the sun.

My brother ran after it, his head thrown back, hoping to catch it before it landed, but the hat fell beyond his reach and its momentum carried it spinning across the ice and into the roosting pool. As it settled in the water he half turned, a reproach forming on his lips, then realised that he was unable to stop.

'Creb . . .!' he called, his arms windmilling as if this would create friction where there was none.

'Creb . . .!'

He pitched forward into the black water.

The lagoon is no more than chest deep, save for one part near the centre where a weighted fishing line drops to the depth of two men. With their instinct for such things, it is that part the geese choose, for there they are furthest from the shore and so least exposed to predators and there, by no coincidence, the water freezes most slowly so that with their numbers, their nocturnal paddling and the small heat that escapes through their thick down, they are able to keep a patch free of ice.

My brother resurfaced as I reached the edge, or as close to it as I dared go. I dropped to my stomach and flung out an arm but he was just out of reach. He stared at me with eyes preternaturally wide as, for a moment, some vestige of buoyancy kept him afloat and he heaved his shoulders and tossed his head from side to side, his mouth working for sounds which did not come. He began to lift an arm, dragging it from the water as if it were lead and I remember, even in that instant, recognising the ambiguity of the gesture – was it one of survival or valediction? Then his sodden clothes and chilled body responded to the greater pull of gravity and before his arm was fully extended he slid back down again. The water closed over his fingertips.

I scrambled up and looked around for help, but the marsh was deserted. There was a soft thud at my feet. His struggling must have propelled him forward for now he was clearly visible beneath me, as if through glass. His legs flapped behind him and one hand scratched feebly at the ice against which his dark hair was flattened. I could see the paleness of scalp at his crown but not, mercifully, his face.

I dropped to my knees again and pounded at the ice with my

fists. It responded with no more than a dull, hollow booming which reverberated away across the lagoon. But he must have heard it, or felt it, for his legs twitched and the scratching of his hand became a brief but frantic clawing. Then, as the freezing water filled his lungs, he gradually went limp and began to slip away into the blackness.

I could, I suppose, have leapt in after him – but then we would both have drowned, since I could not swim either. I could have run back to the village – but somehow I knew, against all the instincts to preserve life, that he was already lost.

In the event, I sat on the ice with my bruised hands and stared at the roosting pool in whose centre, like the waterlogged corpse of some hapless rodent, floated my brother's squirrel-skin hat.

They were strange times then, with stranger still to follow.

In thrall to its tumbledown manor and the poor lands surrounding it, our village lay remote and forgotten beside the winding estuary – a score of low cottages sprawled along the bank, as if they had been spilt there amongst the patchwork of cultivation and little copses. The village was laced by a cat's cradle of narrow, muddy tracks, the most frequented of which led away from the river and across a strip of common grazing where a handful of scrawny cattle and sheep competed apathetically for nourishment. At the far edge of the common stood the well which kept us all in brackish water and served as the focal point, if not the physical centre, of the village. Beyond the well rose a ragged line of elms, whose numbers had been so depleted by successive generations of carpenters, greedy for their fine, hard wood, that the once magnificent village boundary was now marked more by stumps than by trees. As the track passed between them it widened into a lane which climbed gently for a short way, past the cowhouse, the dilapidated church and its overgrown graveyard, through the manorial lands and so to the manor itself, sitting above us on the brow of a grassy eminence like some elderly and censorious relative, with the dense tapestry of the forest at its shoulder.

During his infrequent residences, our lord might have allowed his gaze to stray down across his unkempt gardens and pastures, out over the age-blackened thatch of our cottages – where his eye would not have lingered – and on across an expanse of sluggish, grey water to the curtain of woodland standing in folds along the opposite bank. Glancing upstream, he would have seen – and on a still day, might also have heard – the raucous coming and going of waterfowl amongst the rushes and reeds of the marsh that followed the river's edge to the foot of an escarpment, rising starkly from the water, a mile or so distant. Downstream, his eye would have been greeted by desolate acres of tidal flats, curtailed by the wooded headland which formed the next bend in the estuary. Without turning, he would have sensed the natural barricade that swept around in a great half-circle behind us: first the wall of the escarpment and then the wilderness of forest, rising and dipping with the contours of the land it cloaked, as far as we knew, all the way to the sea.

If his demesne gave him pleasure, we never knew it. It certainly gave us very little. Hemmed in by mud and marsh, cliff and forest, we inhabited as poor and inaccessible a place as nature could have contrived. A single path linked us with the world beyond, running along a small glen that wound through the forest and emerged, six miles on, at a village which for most of us was as distant as the moon.

And now, across the mouth of the glen rose a breastwork of earth, surmounted by a fence of stout timbers hastily erected some three months previously when word had reached us from our absent master that a pestilence was sweeping the land and that we should admit no one from beyond. Nor, on pain of death, should we venture out ourselves until he sent word that it was safe to do so. A single man armed with a crossbow stood guard at the stockade. Within a short while the customary stream of pedlars and packmen had ceased altogether, perhaps because the word had gone round that our village was no longer accessible, perhaps for more sinister reasons – we never knew. Meanwhile, the small craft that plied the estuary became less and less frequent and on the one occasion when a boat attempted to put into our shore, we drove it off with a hail of sticks and stones.

By the outset of that cruel, dark year of thirteen hundred and forty-nine, our isolation was complete and we were enduring the cold season on short rations, growing shorter. We were lordless and also priestless, for the priest had been summoned by his prelate shortly before we received news of the pestilence and had returned a week later to discover that his way was barred. The poor fellow who manned the stockade had stood for some minutes fingering his crossbow while the air reverberated with most unpriestly imprecations, but he had remained obedient to his temporal master and doggedly refused to let the priest in. After a while the priest, who had never shown more than indifference towards his admittedly meagre living, had shrugged, turned his back on the stockade and walked away again – no doubt with a prayer of thanksgiving in his heart.

So yet another straw had come to be laid upon the back of that patient, well-meaning but sadly overburdened creature – our lord's steward. That day, when my father came to him for guidance, his face grew longer than ever. Shocked as we were, we could guess what he was thinking. With the Almighty visiting His wrath so mercilessly on the land, the correct observation of the rites would be more imperative than ever, yet not only was there no one qualified to administer them but the deceased himself was unavailable for burial. A pretty pass, indeed.

He chewed his lip for a while, then disappeared into the manor – to consult the psalter, he said, but we suspected that the real consultation would be with his wife to whom he was known to turn on most matters. In due course he emerged to tell us that although this was a most unorthodox affair, he could, subject to certain conditions, permit a symbolic interment.

Which is how we came to bury my brother's hat.

Towards mid-afternoon, accompanied by the steward and a handful of those villagers who never missed a burial at any cost, my mother, my father and I walked out towards the churchyard. The bitterness had not left the air all day and even the sun, lying low across the river, looked pale and frozen. My father carried a pick, his expression as immobile as the iron with which he was about to prise open the earth. My mother trudged leadenly at his side with bowed head, pinched cheeks and tired, watery eyes fixed on the squirrel-skin parcel – now stiffened with cold – which rested in her mittened hands. She held it awkwardly, a little away from her body; an object of great reverence, yet one she could not wait to be rid of.

I walked behind them, feeling neither grief nor cold; feeling, in fact, nothing at all about anything. The inside of my head was cavernously empty with disbelief. There was a slight ringing in my ears.

The steward halted in the shadow of the yew tree at the furthest corner of the churchyard and explained that he dared not accept responsibility for what might be construed as desecration. The grave, he said, could not be dug within the churchyard. It would have to be here by the tree, where, as he put it, 'The Lord's kindly gaze would surely stray from time to time.' But he was certain, he assured us, that a reinterment in consecrated ground would be permitted when we had a priest once more.

My father heard him out, then without further ado swung his pick at the ground between two roots. As metal met frozen earth a pigeon clattered out of the branches and sped away.

We watched in silence as the pick rose and fell and the effort brought the blood to my father's cheeks. Little by little he hacked away the frost-bound clods and a shallow trench took shape. At length he stood back, laid down the pick and nodded to my mother who walked forward and laid the hat gently on the bare earth. Then she stepped away, dusting her palms very lightly as if to remove some impurity.

For a short while we gazed into the trench and I remember thinking: this is the last I shall see of my brother. But the vacuum of disbelief persisted and that stiff, cold parcel of fur evoked nothing of the lively eighteen-year-old whose head it had warmed. It did not even look pathetic. It was just a hat.

The steward coughed and, without raising his eyes, mumbled something in Latin which sounded more than a little like the grace for our harvest supper. Then, more confidently, he added: 'May The Lord have mercy on his soul.'

'Amen,' we all replied.

My father scraped the clods back into place with his foot and stamped them down as well as the frozen ground would permit. Then he shouldered the pick and turned towards the village again. On the way back my mother's shoulders heaved once and I heard her soft but convulsive intake of breath.

Later that night I came awake to the sound of quiet sobbing and my father's voice whispering gruff consolation. I lay in the darkness for a while, aware that my ears had stopped ringing. But my feelings, if I still possessed any, were trapped, like my brother's body, beneath an impenetrable layer of ice.

I rolled over at cock-crow and for the first time in my life experienced that vague, pre-waking certainty that something was wrong; a misty but pervasive sense of discomfort, almost as if I had been injured during the night. I opened my eyes and saw the mattress beside me. For a moment its emptiness puzzled me. He never woke before I did, already an inveterate lazybones who had to be tickled, prodded, sometimes pummelled to consciousness. But the straw-filled sack had not been slept on. Its coarse woollen blanket was as he would have left it the previous morning . . .

The image of the roosting pool sprang starkly into my mind and I was fully awake. There was a stinging behind my eyes and I found myself willing the tears to come. The unfamiliar, bittersweet pain was so intense that I was afraid I would be consumed unless it found some exodus.

Only once since have I wept as I did that morning and by then I had begun to comprehend the power of grief to cleanse the spirit. But there, with the empty mattress beside me, the grief shocked me more even than the events responsible for it. The sense of desolation seemed so profound that I doubted, no matter how freely the tears flowed, whether I would ever recover. And, in as much as to recover means to recover one's former self, I was right: my brother had taken whatever remained of the child Creb to the bottom of the lagoon with him.

After a long time I felt my mother's hand on my shoulder. She looked at me and nodded but her red-rimmed eyes in the pale, pinched face were so unbearably tender and sad that I turned my head away. Then she reached out and we fell together for comfort. At length, she said: 'The eggs need collecting.'

By the time I came in from the henhouse, I was dry-eyed once more. My mother had swept the floor, set the table neatly and was blowing up the embers in the fire, ready for my father's return with chopped firewood. She was a punctilious woman, despite our scant circumstances, and the cleanliness of her house earned her some respect amongst other like-minded village women. It was a virtue of which my father was justly proud. There were neighbours, no poorer than we, whose cottages were so rank that the stench which drifted from their grimy doorways was almost visible. 'Live like swine, die like swine,' he would mutter, giving them as wide a berth as he could.

Today, as always, he gave an appreciative nod as he entered the room. My mother looked up with a wan smile, then took the logs he gave her and fed them to the fire. As the wood caught and the fire began to roar gently, he stood back with folded arms and looked at me, a curious wistfulness on his usually unexpressive face. I braced myself, but the look faded as he glanced at the bowl on the table.

'Chickens laid well last night.'

'Yes,' I replied.

'We'll have an egg this morning.' He normally deferred to my mother on matters of housekeeping.

'An egg . . . yes.' She covered her surprise by dipping a pan in the stone water jar and placing it on the fire. The eggs, and what he proceeded to do while they were boiling, were the only acknowledgment he ever made of my brother's death from that day on.

'Fetch some water, Creb,' he said. 'The jar's low.'

It was not low. I had filled it myself two days before and I knew well that our normal household needs did not deplete it that quickly. Nonetheless, I fetched the bladders from their peg by the door and made my way to the well, acquiescing readily with his obvious pretext for my absence. I must have been the first there that morning for I had to break the ice before I could haul up the bucket and I remember, despite the bitter cold, feeling a strange inner lightness and warmth as I did so, almost as if I had entered some state of grace. I was surprised to find myself glancing in the direction of the marshes with equanimity, taking in the surrounding landscape of bare trees, frozen fields and ice-rimmed estuary with a pleasantly unfamiliar sense of being at one with my surroundings.

It did not last. As I returned home I saw a plume of smoke rising through the still, pale air and soon realised that it was coming from behind our cottage. I knew at once what it was. My chest began to heave and I stopped at the side of the track, slung about with bulging bladders, as the tears streamed down my face again. Other people were about now and one or two passers-by cast sympathetic glances, but no one attempted to speak to me and I was grateful for it.

I breathed deeply, the air scorching my throat, and walked on to the cottage. My father came round from the back as I reached the door. We entered together and he helped me unsling the bladders. Then, uncharacteristically, he placed an arm around my shoulders and led me to the table. The flames still crackled outside as I ate my egg. I did not need to go into the sleeping-quarters to know what I would see there.

As we finished our meal there was a knock at the door and the reeve stepped inside. He was an unexceptional looking man of early middle-age, neither short nor tall, his complexion somewhat pocked, his eyes a trifle close-set and his belly beginning to droop from the relative indolence of his office. Yet behind the nondescript appearance lay the disposition of a stoat and the word went that he had volunteered his services with the singular intention of fleecing our lord for all he could. We had had our reasons for electing him, nonetheless.

The true wealth of our lord's estate lay in a handful of sandbanks which revealed their presence in no more than a pale discolouration of the mudflats, downstream of the village. For although the sea was several miles away, the estuary was wide, the tide strong and the water did not begin to run truly fresh until a mile or so upstream of the marshes, and during the hotter months, the evaporative power of the sun enabled a goodish crust of salt to gather in shallow depressions scooped from the top of the sandbanks. From these sandy hollows, our lord garnered a substantial revenue. That selfsame salt, however, seeped through the flatlands around the village and it was only with the utmost difficulty that barley or cabbage could be coaxed from the impoverished soil. The only nourishment worth the name was to be found in the uplying acres around the manor and those, of course, were not for the likes of us.

Over the years, the salt had enabled our lord to acquire more pleasant and fecund estates elsewhere and as his absences had grown more and more protracted, a sense of abandonment, then resentment had begun to permeate the village. The gathering of salt was a filthy and backbreaking task which enriched a man we scarcely ever saw and benefited us not one jot.

In these circumstances, our resentment fell naturally upon the steward who kept account of the labour due from us in return for the land on which we built our cottages and the meagre strips of tillage attached to them. As morale gradually diminished and we grew idle, grasping, incipiently mutinous, disputes with the steward became more commonplace and the importance of the reeve, who represented us in such matters, increased proportionately. So when the present fellow offered himself for election, manipulative and unscrupulous to his boots, he came primed to excite our basest instincts and accelerate our moral decline. We elected, in fact, precisely the man we deserved.

I say we, but my father remained one of the few villagers who still gave ungrudging service when called upon to do so and genuinely, if misguidedly, believed that our lot could be improved by simple hard work. He was also one of the very few who openly supported the steward and made no attempt to conceal his dislike of the reeve.

Now he stood up and listened impassively as the reeve, without the slightest acknowledgment of our bereavement, muttered something about having trouble finding men to do a job and told us curtly to accompany him to the cowhouse. That this was not just an excuse to inconvenience my father, whose feelings for the reeve were fully reciprocated, became apparent on our arrival. For there, a very strange thing had happened during the night.

Mounted above the cowhouse roof was an enormous rain-barrel with sufficient capacity to water the cattle during exceptionally hot months when the streams from the forest ran dry and the well was low. Over the years it had begun to leak but its repair was quite beyond the competence of the village cooper and, despite the steward's constant pleas, the money to replace it had still not been forthcoming, with the result that it now dripped continuously, rotting the thatch of the roof directly beneath. In cold weather these drips produced spectacular icicles, for the barrel was supported by a wooden frame that straddled the apex of the roof and around its circumference the escaping water dropped some ten feet or more to the thatch below. In exceptional frosts it was not unknown for the icicles to descend to the roof itself and then, until their tips were melted by the heat from the cattle beneath, it looked as if the great barrel was additionally supported on a palisade of tapering crystal legs – a pretty sight when the sunlight played on them.

The previous night, we learned, the folk who lived on the edge of the village had been woken by a tremendous din from the cowhouse. They had entered to find the beasts stamping and bellowing in panic and one cow lying stone dead in her stall, with her eye hanging out of its socket – through which something had apparently penetrated her brain. Someone had begun to mutter about omens, the pestilence, God's wrath and the like, whereupon a large part of the company had fled home in terror.

A brief inspection by the more stout-hearted had revealed no object lodged in the skull and they had been left to assume that the poor creature had been the victim of some act of vandalism, inflicted with a long, sharp instrument which had been subsequently withdrawn. This had inevitably led to the suggestion of witchcraft and the investigating party had become further depleted.

Now only three or four men remained. In due course one of them had noticed the fast-melting shaft of an unusually large icicle which was missing several inches of its tip. Conclusions had been drawn and shortly confirmed by the presence of a hole which had newly appeared in the rotten thatch above the dead cow's stall. To tired minds, in the small hours, this had seemed explanation enough. But the sceptics had awoken next morning to find themselves chewing on the unpalatable fact that there had been no wind, no thaw and so, apparently, no possible reason for the descent of the icicle. In no time the word had spread and now, before the cowhouse, a sizeable crowd had gathered to speculate in fantastic and morbid detail on the cause and significance of what had taken place, their enthusiasm for the debate exceeded only by their unwillingness to enter the building or, indeed, have any part in the business of clearing it up.

So the reeve had sought out those men in the village whom he knew to be of a more phlegmatic temperament – my father being one – and had assembled a small work party. Some of them were now engaged in the disposal of the dead cow, since it was clear that no one would dare eat it. Others, as evidenced by the sounds of hammering and sawing which came from the steamy interior of the shed, were busy making repairs to the stalls damaged by the panicking beasts. Outside, a ladder led the eye to a roofer who was replacing the rotten thatch through which the icicle had dropped. High above him, clinging to the supporting frame of the rain-barrel, was a lad whose job it was to break off the remaining icicles against a recurrence of the previous night's extraordinary event. This he was doing with obvious relish, snapping the glinting spokes away from the barrel and hurling them clear of the roof like glass javelins so that they shattered, tinkling on the frozen ground beyond.

The crowd parted to let us through and one or two of the bystanders nodded to my father, hastily making the sign of the cross. Whether this was to acknowledge our loss or to protect us from the evils of the cowhouse, I could not say. My father, in any case, ignored them.

'Where are we needed?' he asked the reeve.

'You'd best get inside,' the reeve replied. 'There's a deal of work still to do in there. And you,' he turned to me, 'can lend a hand up there.' He nodded towards the barrel.

I climbed the first ladder to the roof and rested for a moment, breathing in the rich, treacly cattle smell which drifted up in the sharp air. The thatcher, an open-faced young man with a mane of blond hair, looked up from his work and gave me a friendly smile.

'Good day to you, Creb. Going up top?'

Either he had not heard our news or thought it best to dissemble, for there was no trace of condolence in his voice or his eyes.

I nodded.

'Mind where you throw those things, then. Don't want no icicle through *my* head, thank you.'

I set off up the second ladder which was laid against the roof and led to the point at which one of the barrel's large supporting posts emerged through the thatch. I had drawn level with the thatcher when I noticed the hat he had taken off and placed on the roof at his side. It was identical to my brother's. I forced myself on up the ladder, resisting the temptation to let my glance linger on the glistening, russet fur. But by the time I was halfway up the post, hands and feet braced on the pegs driven into it, I began to feel a gathering anger. I could not understand it but it was no less real for all that, nor could I stop it. I loved my brother, truly loved him. He was my one true friend and he had deserted me. He had done it so suddenly, so casually, so . . . callously.

I reached the top of the post and scrambled onto the broad timber frame. Grasping the rim of the barrel, I sidled a little way along then sat down with my legs dangling over the edge. I swung my foot savagely at an icicle. It shattered and a rain of fragments tinkled onto the roof below.

'Hey!' called the thatcher, looking up. 'Careful, I said.' He bent down to the hat and shook the ice from it. This time I could not look away. I found myself thinking of the stiff cold thing under the tree; I pictured it sodden in the roosting pool; I saw it skidding across the ice; and then, the images halted as it cartwheeled across the sun, my brother running towards it, hands outstretched . . .

It was I who had thrown it.

I stood up, nauseous with shock, and stared vacantly at my surroundings, trying to shake away the thoughts that crowded in on me, but the frosted strips of kale, the sheep-dotted grazing, the ice-bound marshes, the tracks and copses around the village seemed merely to sharpen them. I felt the sudden, urgent need to be free of this small, dreary universe. I had been trapped here for twenty years. Nothing of any moment had ever occurred until yesterday – when I had brought about the death of the only person to whom I had every really been close. I found myself eyeing the forest. Did I care that it was infested with cut-throats, wild animals, evil spirits? No. Perhaps the death of the cow *was* an omen.

There was a freezing trickle down the back of my neck. I turned around. The lad had sidled quietly along the timber behind me to slip a handful of ice inside my collar.

'Seen enough yet, Creb? Goin' to stand there moonin' all day?'

'I've seen enough,' I said. It was scarcely my own voice. He gave me an odd look and began to back away.

A moment later I was climbing down the post again. I reached the ground, slid round the back of the cowhouse and set off up the lane.

Two

rowled by named and nameless terrors, the deep forest was as forbidding a place as any of us dared imagine. But familiarity is a powerful exorcist and most of us were perfectly well aquainted with its fringe.

Behind the manor, across the deer park, stood the timber wood – an area of several acres, once ditched and hedged against animals, where ash and alder trees had been coppiced for roofing poles and oaks groomed to grow tall and straight for posts and beams. It had long since run wild but the need for wood persisted. My father, who greatly preferred carpentry to woodsmanship but could still fell a tree cleanly if required to do so, was called here quite frequently and I, possessing no particular skill of my own, tended to accompany him wherever he went. I had always found the timber wood a pleasant enough place, echoing to the companionable cries of our fellow woodsmen, the creak and crash of a falling tree, the thud of axe and adze.

Now it was deserted and silent, but I scarcely noticed as I made my way through and emerged into a spacious woodland of beech and oak where, in autumn, the pigs were driven to feed on mast and acorns. I knew this too, for one of our neighbours was a swineherd and I had often been here as a child. One day I had seen our lord pass by on his return from the hunt with a great fallow stag roped to the back of a horse. It had been messily dispatched with a welter of arrow and spear wounds around its rump and flanks, its throat half torn away by the dogs. I remember noticing the clouded eyes, the grey tongue lolling from its mouth and the flies buzzing about its head, and thinking how ignoble it seemed in death.

The image returned as I scuffed through drifts of frozen leaves. How did my brother look now, his limp body swinging this way and that in the subtle currents of the lagoon? Would his pale, bloated cadaver emerge, like that of the fisherman we had once found on the shore, when the ice melted in spring? His mortal remains had already begun the slow process of disintegration, that much was sure. But what of his soul? I sought comfort in my scant store of celestial imagery: there before him, the majestically-bearded figure of Saint Peter hovered with quill and ledger in hand; beyond the saint, the splendidly-wrought gates towered into the sunlight; beyond them, Creb stretched out on the rack in some dim, flame-shadowed cavern, his bones cracking and his flesh sizzling as the branding iron seared his buttocks and shrieking demons nipped at his privates.

The shock of it halted me. It was an old vision, this one; as frequent an attendant on my childish imagination as the fiends and spirits that stalked the village after dark but one I thought I had cast aside, nonetheless. The last time it had visited me had been three years previously. I had been caught in the act with the ploughwright's daughter – by the priest, of all luck. He lectured us so vividly on the consequences of our iniquity that I had nightmares and was unable to meet the girl's eye for weeks.

But here it was again, with a new and disconcerting potency. Was this the destiny that had called me from the cowhouse? I shivered and turned my thoughts to what else the future might hold for me. I was unskilled and had no special aptitude for anything apart from the employment of a certain natural brawn and so long as my father remained active and continued to need his helpmeet, I had little reason to consider anything else. I had always unquestioningly preferred the company of my brother to that of my peers in the village and now that he was gone I was going to have to make do with my own. There were girls too, of course, but the brief and furtive nature of our liasons, not to mention the fear of their practical consequences, had precluded any possibility of real friendship; and while I had fallen deeply in lust more than once, love had not yet reared its head, nor did I think it likely to while I remained in the village. Like most other people in our small community, I had no learning and was utterly ignorant – even of what lay across the river, or beyond the forest. Yet I had always felt curiously apart, as if I harboured some distinguishing spark – yet to be fanned – which would somehow, one day, release me from the monotony of my present existence. It was this which sent my thoughts wandering away while I was working, which distanced me from the other young men and which, although they could not put a name to it, caused my mother and father gently to despair of me. Only my brother knew of it, because we spoke of everything and he, being more down-to-earth, had not understood but had thought it marvellous all the same because he thought most things I did or said were marvellous. But now he was gone and his absence had started to weigh on my conscience, as well as on my heart.

The air had changed, or the light, or something else in my surroundings. I glanced up to see that a bank of leaden cloud was drifting in over the treetops. With it came a taste of snow. The forest had also altered; now there were hummocks and cliffs, streams and gulleys. The trees were more gnarled and ancient. And moss was everywhere, festooning branches and cloaking boulders, deadening my footsteps. This was no longer the fringe.

I turned for home. A snowflake touched my cheek, then another. There was a narrow ravine before me which I did not recall having crossed before. I sprang at it, missed my footing and a moment later was sitting with a skinned elbow and an aching rump in several inches of freezing water. The rocky sides were steep and darkly glistening with a patina of ice. I stood up but the lip was still some way beyond arm's reach. Upstream, the little canyon appeared to become deeper and narrower and so, with the cold beginning to bite through my damp breeches, I made off in the opposite direction.

By now a wind had got up and the snow was falling heavily, flurrying in the gusts that chased one another down the ravine. Above me the trees were becoming blurred and indistinct and I could sense the sky, very grey and very low, crowding down onto the roof of the forest to discharge itself in a multitude of thick white flakes. I splashed on, blinking continuously, flinching as weird shapes materialised out of the gloom before me and the wind swirled, whispered and moaned around me. This was an unfamiliar wind: a forest wind, laden with cunning and malice, far more menacing than the simple brute that roared through the elms around the village. An animal crashed from a bush above me, dislodging a shower of pebbles which landed at my feet. Something clammy brushed my cheek and before I could raised my hand to it, was whipped away again into the murk. With the hairs bristling at the nape of my neck, I began to run.

A large root had grown downwards through a cleft in the rock. I flung out an arm and hauled myself up, half expecting to feel a hand around my trailing ankles. I stood up, shuddering with relief, and bent forward to catch my breath. Within moments I had begun to shiver. Out of the shelter of the ravine, the wind seemed to sear my bones and the air was so laden with snow that I could see no more than a few paces in any direction. I set off again blindly, knowing only that I had to keep moving.

If He had really meant to punish me, He would have had the pit dug deeper. The thought crossed my mind a moment after I stumbled into the void and found myself, once again, on my arse at the bottom of a hole in the ground. I brushed the snow out of my eyes and glanced around. This was a disused charcoal pit, its turfed roof long gone but still readily identifiable by the hard-baked earthen walls with their coating of greasy blackness staining the snow dislodged by my tumble. A charcoal pit meant charcoal-burners and where there had been charcoal-burners there was more than likely to be a dwelling. I clambered out and peered through the blizzard.

Sure enough, across the small clearing in whose centre lay the pit and several others like it, I could just discern the shape of a hut, heavily mantled in snow and scarcely distinguishable from a large boulder. Skirting the other pits, I made my way across the clearing and walked around the hut looking for the entrance. At first it appeared not to have one, but then I realised that there was, in fact, a small opening concealed by a frame, onto which had been woven a lattice of branches similar to those covering the hut itself. I pulled it aside and ducked in cautiously.

I crouched for a moment in the semi-darkness, conscious only that I was safe from the wind and the snow and whatever else might be abroad with the storm. Then, as relief gave way to exhaustion, I sat down and began shaking my hands in an attempt to restore circulation. But it was not until my eyes were fully acclimatised to the gloom and I was grunting with the pain in my fingertips that I glanced behind me to discover I had company.

He was lying on his side with eyes closed and head pillowed on a leather satchel. Were it not for the stiffness about him, the unambiguous absence of vitality, he could almost have been in repose. I turned around and stared at him for a long time. He was short and slightly built and thin to the point of emaciation. His tattered monk's habit had ridden up to reveal pale, skinny, almost hairless shanks. Beneath his tonsure was a high, scholar's forehead, an undistinguished nose and a lightly receding chin from which sprouted a meagre stubble. Although robbed now of his expression, I imagined him to have worn a mildly questioning look, exaggerated, perhaps, by short-sightedness. He was not very old – five years my senior at the most – and death must have come to him through starvation or cold or, most likely, a combination of the two. But what had brought him here? The forest was no more a place for defenceless clerics than for villagers like me. And why, in all of this unhallowed wilderness, should I have found myself in the one place containing a dead monk? If destiny, or the Almighty, had any hand in what was happening to me today, it was becoming less comprehensible by the moment.

I stood up and stamped my feet. Outside the blizzard raged on and snow drifted continuously through chinks in the walls and roof. I had never felt colder in my life.

In the centre of the floor was the remains of a small fire and it occurred to me now, from the way my companion was lying with his body slightly curled towards it, that it might have been of his making. If this were so then he would have a flint in his satchel. I bent down beside him, tugged at the satchel and encountered a problem. He had the shoulder thong wrapped around his forearm and clenched in his fist, so close to the mouth of the satchel that it was drawn tight and I could not slip my hand inside it. Nor could I prise open his fingers to release it. With a knife I could simply have slit the thong and pulled the satchel free. But I had no knife. I wrestled for a moment with my conscience, then muttered 'Forgive me, Brother,' grasped the little finger and pulled sharply upwards. There was a brittle snapping and the finger stood up like a miniature billhook, at right angles to the back of his hand. There it remained, the waxy flesh still indented from my grip. The other fingers followed in a similar fashion and I was able to pull the satchel free. I carried it to the entrance, where there was a little more light and upended it. The contents were most curious but did include a flint, as well as the remains of a very old loaf and a small portion of rock-hard cheese. I lit a small fire with the driest stuff I could prise from the walls of the hut and then, gnawing on the cheese, squatted down to examine the rest of my companion's effects.

There was a small knife and a queer lump of stone – pale greyish, peppered with tiny holes and light as a feather. There were three goose quills and a small lidded inkhorn. And there was what looked like a goat's tooth. The satchel, however, was not quite empty. I thrust my hand into it and pulled out a sheaf of parchment, bound with a ribbon, that had lodged at the bottom. I could no more read than play the fiddle, but curiosity compelled me to untie the ribbon and roll the parchment flat.

There were several sheets, most of which were densely covered with writing, interspered with figures and symbols. Although its efforts were quite indecipherable to me, the hand itself was neat and the ranks of clean, angled letters were pleasant to look at, in an orderly sort of way. The figures, equally precise, depicted a variety of objects, some of them familiar and some of them unfamiliar, but their significance, if indeed there was any, entirely eluded me. Here there was a crowned sun and here a fish; there a moon and attendant stars; here a phial or jar; there a dragon; there a pair of bellows – and elsewhere, many configurations of straight lines connecting groups of unknown symbols. Whatever it all meant, I had no doubt that it resulted from years of monkish reasoning, as far beyond my attainment as anything I could imagine. I glanced at my companion, oblivious to the thumbing of his learned tract by an ignoramus, and wondered whether it could have been his own work.

I continued idly to turn the sheets and my mind began to wander: picturing the devoted monk at prayer in his monastery; imagining him seated at a table with writing tools laid out; wondering what great and marvellous knowledge it was that flowed from head to hand and thence to the parchment.

The final page caught my attention. Quite unlike the others, it contained no writing. Instead, it was marked down the centre, from top to bottom, with a narrow pair of parallel lines broken at various points by minute drawings. I squinted at it in the meagre firelight and was able at length to distinguish what might have been a castle, a gallows, a church and steeple, an anvil, a well, a cave . . . Dotted around them were tiny trees and animals and other markings I could not identify.

I gazed at it in fascinated incomprehension. It spoke to me in some way that the other crowded pages did not: here the drawings held the key and these I could largely understand, even if their purpose remained unclear. There was knowledge of a different kind here, I was sure of it – not, perhaps, the answers to such questions as why the icicle had fallen, or whether God meant to punish me, or whether I would ever find, or even want, a wife – but something else of very great value which I suspected to be far less remote from my grasp than the knowledge contained in the writing. For the first time in my life I deeply regretted my ignorance.

I folded the last sheet carefully and placed it inside my jerkin, then gathered up the others and tied them with the ribbon again. After a moment's reflection, I pocketed the flint and knife and replaced the other writing tools and the parchment in the satchel. The satchel itself I respectfully returned to its place beneath my companion's head which, temporarily bereft of its pillow but still supported by the rigid neck, gave the eerie impression of being cushioned on thin air.

I do not know how long I spent in the hut, but the fire had gone out and I noticed now that it had become quiet outside. I went to the entrance, pulled aside the covering and glanced out. The wind had dropped and through the few fine flakes that continued to fall I could see shifting cloud and vague patches of pale blue.

Stiff and chilled, I set off through the silent, snowbound forest.

Although I had wandered no more than three or four miles before the blizzard had come, it took me some time to find my way home. The features of the forest seemed to have altered beyond recognition and on several occasions I found myself crossing my own tracks. But it was now a much less alarming place. It felt spacious, light, purged by the snow.

In due course the sky cleared and I was able to take my bearings from a westering sun which cast long bluish shadows across the powdered woodland floor. I felt quickened again by the squeak of my boots in the fresh, dry snow and the sharpness of the returning frost. I had been into the forest and had survived – no, not survived, for I had never been threatened. I had neither seen nor heard any trace of the horrors that were supposed to have beset me. Wind and snow and holes in the ground, those had been my only foes. But I had achieved a kind of freedom. And I had discovered a treasure – a key, I sensed, to something far greater than the universe of my present acquaintance which now seemed more ramshackle, more moribund than ever. For my brother's sake, I would learn how to use that key, at whatever cost.

I returned home at sunset. My mother was at the fire, busy with her cooking. She glanced up and for a moment her look softened with relief and curiosity, but she did not say anything. My father sat at the table whittling a piece of wood. Without lifting his head, he said:

'Where d'you get to?'

'Nowhere. I just wanted to be alone. Got caught in the snow.'

He nodded and continued whittling.

'Shouldn't leave a job half done, y'know. Specially when it's reeve's work.'

'No. I'm sorry.'

He grunted, then looked across to my mother and asked her how the meal was coming along. As she replied, I could not help noticing the contrast in their eyes. For hers, although tired, were still lit with the spark of response, while my father's now seemed slightly clouded, as if in echo of some deadening inner pain.

Later that evening I went to bed, turned my face away from the empty space beside me and instantly fell asleep. During the night I dreamt of charcoal-burning monks with crooked, parchment fingers.

Three

he previous day's events were clear in my mind the moment I awoke and with them came an unprovoked and entirely unwanted thought: it was neither cold nor hunger that had dispatched the monk. It was pestilence. He had brought it with him from wherever he had come and now the menace, lurking these months on our threshold, had chosen me to effect its silent and unobtrusive passage into our midst.

If punishment were in the air, this was its ultimate refinement. What should I do? Throw myself into the river and join my brother? Flee the village? I did not feel ill, but my heart was beating unusually fast and my palms were cold and damp . . .

I rose and dressed and went out to seek reassurance in the ritual collection of the night's layings. In due course the familiar pungency of the hut, the soft clucking of the hens in the gloom around me, the groping through straw for the warm brown eggs, began to have their effect and as I grew a little calmer it began to occur to me that perhaps I might profit from doing nothing for the time being. Wait a day or two to see if any aches or pains developed, any spots or swellings. It would be an unpleasant wait but at least I had my new discovery for distraction. I let the hens out and walked down our snowbound strip of tillage to a tree-stump, where in milder weather I often sat to look out over the river. I pulled the parchment from my jerkin and examined it again.

What was it that appealed to me so much about something I did not understand? There was little ambiguity, I guessed, about the images themselves – they were simply what they appeared to be; a castle, gallows, steeple and so on. But why were they there? What was their collective significance? They were all directly linked by the lines as if by a track, or road. Perhaps, then, this was a drawing of some part of the countryside. If it were, it was an ordered and tidy place where the road ran straight, the land was flat and all but the essential detail had been stripped away – an immeasurable improvement upon the crowded and confusing landscape through which I had blundered yesterday, for instance. But even if it were the drawing of a place, real or imagined, the perspective was very strange indeed. My dead companion must have been imagining himself to be a bird when he drew it.

'Creb! *Creb!*' My mother was gesticulating.

I returned to the cottage, picking up the basket of eggs on my way and went inside.

'What is it?'

'We must go. We're all called to the manor.'

I was about to ask why when the answer came to me, but my mother was damping down the fire and my father was heaving on his coat and neither of them saw my face go slack.

Pain of death. Those were the terms, so we had been told, upon which we disobeyed our lord's injunction to remain in the village until the pestilence had passed. Someone must have seen me going into the forest, or coming back from it. No, not *me*, but my footprints! From the far corner of the desmesne, where I had eventually emerged from the trees, they would have run across the pasture, through the outskirts of the village and straight to the door of our cottage. They would have been like beacons in the fresh snow yesterday evening. They still would be today, after last night's frost.

I kept my head down as we made our way through the village, avoiding the eyes of the other folk who exchanged greetings and gossip as they joined the procession. I felt my bladder contracting. By the time we were halfway up the lane I was in dire need of a piss. But I dared not do anything to make myself conspicuous.

We reached the manor and filed slowly into the hall where the chatter was now muted with anticipation. Pressed up against the other bodies in the throng I began to feel that all eyes must be upon me, that I must be reeking of sickness. Someone nudged me. There was a rustle under my jerkin and I remembered that I was carrying with me the most incriminating evidence of all.

Although the hearth was empty, the grand, raftered chamber

was already warm and steamy as a hundred or so bodies jostled for a view of the low dais at the far end, upon which now, looking distinctly ill-at-ease, sat the steward. At his shoulder stood the reeve, his eyes flickering back and forth across the gathering.

Presently the steward stood up, raised his hands for silence and the chatter died away. He was a tall, stooping man with pale, soft features, almost effeminate beneath the fur brim of his hat; thus as marked from the reeve in his appearance as in his temperament and loyalties. A scrupulous administrator and meticulous registrar of his master's affairs, he would have been better suited to the post of secretary to a bishop or abbot than that of agent to an absentee lord, and it was clear that of all the authority he had recently found thrust upon him, this was the most unwelcome yet. The reeve, by contrast, held his head forward and cocked slightly upwards, as if he were scenting the air for yet another opportunity to further his own ends.

Arriving amongst the latecomers, we had taken our place near the back of the hall. Now I glanced at the door, a couple of paces from where we stood, and prayed fervently that the steward would remain in control of the proceedings.

The steward fiddled nervously with his sleeve as he scanned the expectant faces before him. Then, at length, he cleared his throat and said: 'Friends, it is three months since our lord so providentially sent word of the pestilence, three months since we sealed ourselves from the outer world – and I offer you my esteem, every one of you, for the courage and endurance with which you have faced such adversity. I have prayed each day, as I believe we all have, that we would be spared this dread affliction and it would seem that the Good Lord has heard our prayers. But now, although we survive the depredations beyond, we have been . . . smitten in our midst.'

He paused, as if deliberately prolonging my agony, then continued: 'I bring you the regrettable news that a murder has been committed . . .'

In all the uproar that followed I felt myself quite alone, almost disembodied, as my mind grappled with this startling turn of events. Not pestilence. Not criminal egress. But murder. Well, there was my brother – should anyone choose to make that interpretation. And, if fate was particularly against me, there could even be the monk. I could easily have been followed to the forest and I would be equally hard pressed to exculpate myself there, especially after what I had done with his fingers. The gibbet appeared before me. I eyed it with morbid fascination.

With some difficulty the steward restored silence and now an altogether different mood descended on the hall. It was one of almost lecherous expectation, a deep hush born not of respect for the steward but of a desire to miss not a word, not a detail of what was to follow.

'A murder has been committed, friends,' he reiterated, 'and an accusation has been made.' A murmur arose again but died quickly as he went on: 'And now, since these are troubled times and the . . . er . . . proper authorities are not available to us, so we must . . . um . . . try to reach the bottom of this . . . ourselves.' He ran a hand across his forehead. 'We are fortunate, however, that it is our reeve himself who discovered the . . . er . . . felony and so I call upon him to summon the accused and lay the charge. Master Reeve . . .' He sat down and began an intent inspection of his fingernails.

Now the reeve came forward. I held my breath as he stepped briskly from the dais and the crowd parted before him. But it soon became apparent that he was making not for the rear of the hall where I stood, but for a smaller doorway halfway down one side, through which he shortly disappeared. I had a moment to collect my thoughts and conclude that he had gone to fetch some piece of evidence – the disinterred hat perhaps, or the satchel – when he reappeared with his hand firmly on the shoulder of the wheelwright, a short, barrel-chested man whose saturnine features were knitted in an expression of sullen indignation.

I watched incredulously as the reeve marched his captive towards the dais and there spun the man sharply to face us. The wheelwright stood glowering, his broad shoulders squared by the thongs binding his wrists behind his back, while the reeve declared: 'This is the fellow. And I hereby accuse him that last night he did stab with his knife and so did foully murder one Hubert, mole-catcher and bachelor of this parish, now deceased. What say you, Master Wheelwright?'

The wheelwright scowled at his accuser. 'I never done nothin' of the sort, Master Reeve. That's the truth – and you know it.'

The reeve glanced towards the steward, still intent on his manicure, and said loudly: 'The accused denies the charge.'

'Ah . . .' replied the steward. He looked up without enthusiasm. 'Very well then, present your evidence.'

The reeve responded with a quite unnecessary bow, then began: 'Some hours ago, when it was still dark, I heard a noise of cattle. After what happened the night before last I thought I should see if anything was amiss. I left my bed and took the shortest way from my dwelling to the cowhouse. The sky was clear and snow on the ground gave a fair light to see my path. I was close by the salt-store when I heard struggling and oaths so I ran forward. As I came round the front of the store I found this fellow and the deceased, God rest his soul, locked in mortal combat.'

At this, the wheelwright's scowl deepened and he started to say something but the reeve ignored him and continued: 'I went to pull them apart and as I drew near, the accused put his knife in the deceased's ribs and he fell down the steps to the salt-store door. Then I thought only to apprehend the felon so I jumped at him and wrestled with him and with much hard fight I took his knife, but not . . .' he paused dramatically and raised his arm so that the sleeve fell back to reveal an ugly gash, '. . . but not before he had done this.'

There was a sympathetic muttering from the crowd. The wheelwright shook his head in disbelief. I watched and listened in a curiously suspended state, still only half crediting what was taking place before me.

'And so I overpowered him and brought him back to the manor,' concluded the reeve, lowering his arm again, 'and now he stands accused for all to see.'

'Thank you, Master Reeve,' said the steward at length. 'A very . . . er . . . lucid description of events.' He gave a deep sigh, then continued: 'Now, we have a victim, an accused, a witness's account . . . can you perhaps suggest a motive for this affair, a reason for their quarrel?'

Without a word the reeve reached behind the steward's chair and produced a leather bag into which he thrust his hand and withdrew it again with a flourish. Extending his arm towards us, he flattened his palm to reveal a quantity of salt. He paused to let this evidence have its effect, then said: 'This bag was upon the accused's person as they struggled. I say that he had robbed the salt-store and the deceased found him and tried to apprehend him for a thief and was murdered for his pains.'

'But the salt-store is locked, is it not? And I have the key.' The steward sounded confused. 'How would the wheelwright have come by it?'

The reeve shrugged. 'He had no need of it. He has tools that would do for a key. And the door is old and the lock is worn, as you know, Master Steward,' he added pointedly.

'Yes, that is true.' The steward gave a weary nod. 'So, the wheelwright gains entry to the store and helps himself to salt. As he emerges, the mole-catcher, who happens to be passing by . . . but why would he be passing by? Not setting traps, surely, in the middle of the night . . . and at this time of year?'

There was a general titter but the reeve interjected sanctimoniously: 'That we shall never know.'

'No,' agreed the steward, 'we shall not, alas.' He picked up his thread again. 'So the mole-catcher who, for reasons unknown, happens to be passing by, catches the wheelwright red-handed and attempts to apprehend him – at which point you, Master Reeve, on your way to the cowhouse, discover them locked in combat and, before you can intervene, see the mole-catcher stabbed to death. You grapple with the wheelwright and eventually overpower him and . . . well . . . here we are.' With a pained expression, he turned to the accused. 'And you, Master Wheelwright – since you deny this charge you no doubt have a somewhat . . . er . . . different version of events. Be so good as to let us hear it, please.'

'Not till I'm undone,' muttered the wheelwright, still scowling.

'I beg your pardon?' The steward cupped a hand to his ear.

'Not till I'm undone,' repeated the wheelwright, raising his voice and wriggling his hands behind his back. 'My bonds, Master Steward. I'll not speak like a trussed fowl.'

'Oh, very well. Master Reeve . . .'

The reeve nodded sourly and undid the thongs. The wheelwright rubbed his wrists in silence for a while, then wiped his mouth with the back of his hand.

'What you just heard is a pack o' lies, Master Steward. Believe it if you want, and I 'spect you will, but as God's my witness, what I tell you now's the truth.' He glanced venomously at the reeve, then continued: 'I was awake with me wife's coughin', see, and as I went to get water I heard somethin' outside. But it weren't cattle noise. It were footsteps, crunchin' on snow. I looks out and there's reeve goin' up lane. Up to no good as usual, I says to meself. I fetches me wife water then I slips out after him, curious, see. On the way, there's old Hubert awake with the pains in his bones. He hears us passin' and he comes out too and he don't trust reeve no more'n I do. So we follows him up lane, toward cowhouse. When he gets to salt-store, in he goes and we waits for him and a few minutes later out he comes again, bag on his shoulder. Then Hubert steps out to challenge him, see, and reeve has his knife out in a flash and he sticks the poor fellow in the ribs. Then I'm in a red rage 'cos old Hubert never done no harm, and I jumps out and fights with reeve and gets his knife and cuts his arm but he makes me drop knife in the snow and I slips and knocks me head on a stone – and next thing I know he's got me so I can't move and he's tyin' me up and . . . why he didn't do for me too, I dunno.' He lapsed into a melancholy silence.

The reeve, who had listened without the slightest reaction, now turned to the steward and said: 'An incredible tale, think you not, Master Steward?'

It was evident that the steward now wished he was somewhere else altogether. For some moments he stared wistfully at his feet. Then with a brief, conciliatory glance at the reeve, he turned to the wheelwright. But the fleeting exchange had not escaped the wheelwright. Correctly interpreting it as his doom, he became suddenly galvanised and, before the steward could say a word, began to harangue the crowd: 'I'm unfairly accused and you know it, all o' you . . . he's lyin' to save his skin . . . been stealin' salt these past months . . . featherin' his nest . . . that's why he put hisself up . . . and now he's done for old Hubert . . . it was him all right . . . I never killed nobody . . .' He was waving his fists furiously in the air – 'Come on, you shithearts . . . pissbrains . . . waterloins . . . won't no one vouchsafe us?'

Beside me, I sensed my father stiffen, saw his jaw lift. I waited for him to step forward, or speak up, but then the moment seemed to pass and his shoulders drooped as if he were overcome with lassitude. I caught my mother's eye and she looked away.

There was a sudden commotion up ahead as someone began to push through the crowd, making for the door. As they drew closer I saw that it was the steward's son, a lad of about my brother's age who was seldom seen around the village on account, so we were told, of his delicate health – although gossip would have had us believe that there were other, more sinister reasons.

When he was level with me he stopped abruptly, almost as if reined in from behind, and spun round to face the dais again. Our immediate neighbours edged back and I found myself doing likewise, still seeking instinctively to escape attention. It was not until the mixed glances of deference and distaste prompted me to study him more closely that I realised he was flushed and breathless and in a quite extraordinary state of agitation.

Up on the dais, the wheelwright continued to rant and wave his fists. The steward, with an increasingly dismal look of resignation, was doing his best to listen to whatever it was that the reeve was whispering into his ear and it was the reeve upon whom the lad's dilated eyes were now fixed.

Finally the reeve stood back and roared: 'Pray silence for the steward!' The steward rose to his feet and the wheelwright, seeing the situation was hopeless, waved a final despairing fist at the crowd then fell silent.

'Friends,' said the steward, 'since no one will come forward on Master Wheelwright's behalf and I can therefore conclude that we have now heard all the evidence, we must decide between his word and that of Master Reeve. In this I must seek your assistance, since it is a matter to be decided not by me alone but by all of us here present. Do I have your assent?'

There was a general muttering and nodding of heads.

'Good. Now, all those who favour Master Wheelwright, make themselves known.'

Scattered grunts followed.

'All those who favour Master . . .'

'No!' It was the steward's son, his voice thin, almost strangulated. There was a collective gasp and every head in the hall turned towards us. He was now trembling. The flush had drained from his cheeks and he was grinding his teeth with the apparent effort of remaining erect.

'He's lying,' he called. 'The reeve's ly . . .' His eyes rolled suddenly upwards and with an involuntary cry he fell to the floor, his limbs locked in spasm. Decorum abandoned, the bulk of the gathering pressed noisily forward to see what was going on, only to run up against those around us who were trying to back away still further, their alarm quite apparent now.

I remained where I was, rooted in horrified fascination as the lad's legs and arms began to twitch convulsively and a bloody froth bubbled through his clenched teeth. A moment later his Adam's apple began to work furiously up and down and his jaws from side to side as if he were trying to gulp something down with his mouth closed. A bluish tinge spread around his lips and it came to me then that he was swallowing his tongue.

I found myself dropping to my knees at his side and trying to prise open the jaws but they seemed to have closed on steel hinges and, as his limbs continued to jerk, it was all I could do to hold his head still. There was a further commotion as the steward pushed through the gawping throng and knelt beside me, his face racked with concern. Together we succeeded in opening the lad's mouth enough for his father to insert thumb and forefinger and fish for a purchase on the quivering, fleshy organ, slippery with blood and saliva, which was curled backwards and already halfway down his throat. At length he was able to hook his finger around it and roll it forward. I fumbled in my pocket and pulled out the monk's knife, waited till the steward's hand gave me the necessary space, then wedged it as far back as I could between the lad's jaws. The gulping stopped and the steward moved round and held his son's head on his knees. After a moment or two, the convulsions became slower, more irregular, and eventually died away. Within a short while a state of deep relaxation seemed to take hold. The steward lifted the limp body from the floor, muttered a word of thanks to me and carried the lad from the hall.

For some time people hovered uncertainly, talking only in whispers. Had the proceedings been abandoned? It appeared so. The reeve had gone and so had the wheelwright. But there was still some reluctance to move. Then a child who had been standing near us sniffed loudly, wrinkled its nose in disgust and pointed to where the lad had lain: 'Ugh!' it announced. 'He shat his breeches!'

This observation earned it a sharp cuff on the ear and it began to wail. All at once, normal conversation was restored and we started to file slowly from the manor. It was snowing heavily again.

Not until later that day, after the mole-catcher had been laid to rest, did it become known that the wheelwright had vanished. It appeared that as the crowd had surged towards the back of the hall, he had taken advantage of the confusion to elbow the reeve in the stomach, grab the bag of salt and bolt for the side door. Doubled up from the blow, the reeve had taken some moments to regain his wind, by which time the wheelwright had dived out of a window and was heading like a hare for the forest.

Now the reeve had retired to hold court in his dwelling, where he could be heard inveighing loudly and furiously against the steward for allowing him to be publicly discountenanced and thwarted in his attempts to see justice done. But the truth of the matter was that the morning's events had at last furnished him with precisely what he had been waiting for: a pretext upon which to commence in earnest his campaign against the man whom he had long regarded as the only obstacle to his wholesale pillage of the estate.

The rest of the village, meanwhile, was agog with talk of the steward's son. Although such a thing had not occurred in recent memory, fits were not unheard of and in less troublesome times would soon have been forgotten. But in the prevailing climate of fear, superstition and mutual distrust, the fact that the seizure had occurred at a public gathering and, furthermore, that the utterance preceding it was generally recognised to have been accurate, had resulted in it making a considerable impression. Now it was rumoured amongst the majority of villagers that the steward's son was pestilence-ridden, or the possessor of terrifying supernatural powers, or both. The more sanguine minority, most of whom were secretly delighted at the reeve's discomfiture, remained content with the notion that the lad was his nemesis, sent by God.

That evening, however, my own preoccupations were elsewhere. I seemed to be enjoying some curious immunity from what I still suspected to be my just deserts. There was, of course, still the question of whether or not I had contracted the pestilence. But should there have been some divine plan to bring me to book it was, on the evidence so far, no more than a catalogue of missed opportunities. I began to wonder whether I had misinterpreted all that had taken place during the last few days. Was there, in fact, some perfectly coherent chain of events, set in motion by my brother's death and deliberately constructed to guide me to the monk and his strange little drawing? Could it even – I indulged myself briefly with the proposition – have been arranged to console me for my loss? No. A dangerous train of thought to pursue. Yet however obscure its purpose might be, the very possibility of such an invisible order lent force to my conviction that, barring pestilence, I should in due course unlock the knowledge the little drawing contained and profit by it . . .

I had thought little more of my intervention in the hall until, at supper, my mother said proudly: 'That was a fine thing you did today, Creb.'

I shrugged and mumbled something self-deprecating.

'Well, I think it was a good, Christian deed, whatever you say. And the steward'll not forget it, though Lord knows, the poor soul's got troubles enough for all of us.' She glanced at my father. 'He'll think well of Creb now, won't he?'

But my father had been lost for some time in morose contemplation of the fire and as I looked across at him I felt my own sorrow, temporarily forgotten, beginning to well again.

Four

week after the trial of the wheelwright, I was still as sound in wind and limb as I had ever been and, while the pain of my brother's absence continued to gnaw, my fear of – or perhaps it had been a desire for – pestilence or indeed any other form of retribution, had begun to recede. Fate, however, still had one definitive twist in store for me.

The reeve was at our door early one morning, requiring our labour again for repairs. This time the cause was more mundane, for when there was snow on the ground the sheep were fed with hay, but supplies were now running very low. The previous day, after swiftly consuming their meagre allowance, the hungry animals had gone off in search of alternative fare, knocking down a fence and making short work of a small kale of field.

As he spoke briefly to my father, the reeve seemed no more or less uncivil than usual, but catching sight of me through the open door his eyes narrowed disconcertingly.

My father and I made our way to the field and joined the couple of other men who had already begun to re-erect the brushwood fence. There we we set to work, digging holes for new posts. I was holding one of these posts in position, ready for the swing of my father's mallet, when a small boy approached us.

'Steward wants to see you,' he announced, breathless with the importance of his mission. My father laid down the mallet and put on his coat.

'Clear that up while I'm gone,' he said, pointing at the remains of the old fence. 'We'll burn it later.' He started to walk away when the boy said: 'Not just you. Wants both of you.'

My father stopped and turned round. 'Come on, then.'

I released the post and it fell to the ground. I had the sudden presentiment that this was something to do with the reeve, related in some unpleasant way to the trial or to the events preceding it. My hands trembled as I slipped on my jerkin but my father said nothing and we walked up the lane in silence. The boy accompanied us and as we reached the manor, pointed up towards the steward's living-quarters above the storerooms on the opposite side of the courtyard to the main hall.

This time my father raised an eyebrow. Audience with the steward invariably took place in his booth just inside the entrance to the storerooms. So this was not a village matter.

We entered the first one, musty with grain and climbed the steps to the steward's door. My father knocked respectfully. Despite the digging and the brisk walk, I found myself shivering. After a moment the steward appeared.

'Ah, come in, come in.' Although his voice sounded tired, the greeting was measured. It could have been the prelude to almost anything. He ushered us inside, then gestured to a curtained recess at the far end of the room. 'Come with me,' he said to my father. 'Creb, you wait here.'

He drew the curtain behind them and I was left alone in the long, low attic chamber, wishing that I had never gone down to the marshes that morning. How long ago had it been – eight, nine, ten days? I could no longer remember. My life seemed to have become accelerated, like that of a small animal or insect, and it was beginning to complicate my sense of reality. I found myself studying the dust motes which swam lazily through a shaft of sunlight admitted by a chink in the timbers, envying their apparent aimlessness. The warmth of a good fire, the murmur of voices behind the curtain, occasional movements from the sleeping-quarters beyond, the muted clump of feet in the storerooms below . . . the atmosphere was almost soporific . . . if only my mind would be still . . .

The steward and my father emerged from behind the curtain. My father's expression was as impassive as ever but the steward seemed to have undergone some subtle easing of tension, as if recently relieved of a headache.

'I prayed you would support me in this,' he said. 'I am most grateful.'

My father merely nodded and made his way to the door as the

steward led me to the fireside and indicated a stool. I sat down and waited as he stood looking at me, silently and intently.

'Well, Creb,' he said at last, 'what do you suppose life holds for you, here in the village?'

I was too startled to respond. I looked at him blankly. He smiled a little and broached the question a different way: 'Do you enjoy your work?'

I shrugged. The idea of enjoying work was not one that I had previously entertained. 'Don't know sir,' I replied.

'Your father is a fine craftsman, Creb, when he has the chance – one of the best in the village. He is an honest man, too. He knows what it takes to make a good carpenter and he admits that his son does not have that quality.' He paused, looking at me directly. 'He says you dream. Would you agree?'

I glanced down at my hands which already bore ample testimony to misdirected hammer blows and careless saw strokes. 'Maybe,' I replied hesitantly. Surely I had not been called here to be reprimanded for lack of concentration?

'But is it just dreaming, Creb, or is it really thinking?'

He allowed me a moment to wrestle with the distinction, then intervened: 'A fine point, Creb, but an important one. And I shall answer it for you. A dreamer has a vagrant mind, a wilful mind which leads him wherever it wishes. A thinker, on the other hand, controls the direction of his mind so that it takes him wherever *he* wishes. Do you follow me?'

I nodded uncertainly.

'Now a dreamer would have stood by as my son lay on the ground. But a thinker would have acted as you did. A thinker would have placed his knife in the lad's mouth because he was capable of realising, even though the experience was new to him, that it would halt the swallowing.'

The steward paused for a moment and I found myself wondering whether this was some veiled rebuke or a genuine compliment.

'You are a thinker, Creb,' he continued, 'and I need a thinker – or rather my son does, because sometimes he is unable to think himself and he needs someone else to do it for him. Someone who is ready to look after him. Someone who will not be troubled if he says – strange things.'

'Like . . .' I began, then bit my tongue.

But the steward nodded. 'Indeed. As he did the other day. I

should not have let him attend the trial but he begged me. He is so . . . interested in everything. And then, of course, he became agitated by the proceedings. You see, Creb,' he went on, 'much of the time Roland is a normal, wholesome lad, as eager to discover and learn and play the adventurer as any, and it is wrong for him to have to be cooped up here in the manor like a capon. He should be able to lead his life like other lads – indeed he deserves it. But we do not know when his seizures impend and by the time he is aware of it, it is often too late.'

He sighed and looked into the fire. There was pain in his eyes, or perhaps it was remorse or regret, I could not be certain; nor did I know whether to feel flattered or discomforted at being admitted to the confidence of this soft-spoken, elderly man who was so manifestly my superior in every respect.

'I cannot devote the time to him that I should like to,' he went on. 'He has no brothers or sisters and he should no longer be spending his days with his mother at women's tasks. The sum of it is that his existence could be so very much more rewarding if only he had the freedom to move, to experience the world beyond these walls, to observe and enquire. But until now I had dared not believe that there was anyone of a suitable age amongst us who would understand, or even if they did not understand, would at least not be – afraid.'

He paused again and looked at me directly. 'What I ask, Creb, is this: would you become his companion?' His voice contained an echo of hope and I knew at once that this was more than merely a tacit command to an obedient and unquestioning serf. My agreement in this matter, spontaneous, uncoerced, was of great import- ance to him – and as I recalled the looks on the faces in the hall at the wheelwright's trial, I began to understand the reason for it.

'What do you say?' he asked.

'Well, sir – there's my work,' I replied. 'I know I'm not much skilled at anything but there's our land to tend too, and . . . I'm the only one now.'

'A sad, sad loss for you, I know.' His eyes creased with sympathy. 'But I have discussed it with your father and we have – come to an arrangement. He and your mother will not be disadvantaged by your absence, Creb, I give you my word.'

If anything advanced the theory that my life was now spinning out of control, it was this. Well let it spin, I thought. This was clearly not the time to start weighing consequences.

'I'll take care of him,' I said.

He smiled. 'Good. Then, you will start tomorrow morning. Come along when you have eaten and I will decide how you can profitably pass the day. Do you have any questions?'

I hesitated before asking: 'What should I do if – it happens again?'

The steward nodded again. 'Fetch him back here if you are able. If not, do just as you did in the hall. He is in no real danger as long as there is someone there to prevent him from harming himself. And it passes swiftly.'

I rose from the stool.

'Thank you, Creb,' he said, putting a hand on my shoulder as he led me across the room. 'You are an intelligent young man. Stay a thinker, not a dreamer, and you will do what I have asked of you well.' He began to open the door, then paused and lowered his voice. 'Roland has knowledge of these seizures, of course, but he has no recollection of what he has said or done at those moments. For his sake, I should like it to remain that way for as long as possible.

I nodded, then closed the door and made my way slowly downstairs.

My mother fussed over me next morning as I finished my breakfast and prepared to leave. Whatever any of the villagers thought privately of the steward, his position continued to engender a certain deference in all of us and now my mother's son was to be his son's companion. I think she saw my new duty as an honour but a well-deserved one for all that. She smiled proudly and leant forward to brush a speck of mud from my jerkin. 'Look after him.'

It was a sparkling morning and bitterly cold again. During the snowfalls the passage of people and animals had created puddles in the ruts of the lanes. Now they were coated with ice, milky and whorled, which cracked sharply as I stamped on them. Frosted cottages and trees stood clearly etched against a brilliant sky and, with its snow-covered roof and the faintly silvered wall of the forest at its shoulder, the manor – for once – appeared almost benign.

I reached the storerooms to find the steward already seated in his booth, examining a ledger. He gave me a genial 'Good-day' as I entered and asked me to fetch Roland so that he could tell us what he had in mind for the day. I climbed the stairs and knocked at the door and a moment later his wife appeared. She was an ample woman with greying hair drawn sharply back from her temples. Her features were broad and strong and dominated by a pair of lively eyes from which, I guessed, little remained hidden.

'Creb,' she said, drying her reddened hands on her apron. 'Come in. Roland is waiting.'

'Am I – am I late?'

'No, no. But he has been up since dawn.' She chuckled affectionately. 'It is something quite new for him, you understand – a friend.'

The word grated. I had not considered the possibility of my new duty entailing friendship. I followed her into the chamber and glanced across at the lad who sat by the fire, already muffled for the outdoor chill. He looked up with a tentative smile which I did not return.

'Roland, this is Creb,' said his mother.

'I know,' said Roland. 'You helped me.'

I shrugged gracelessly and mumbled that it was nothing.

'Oh, it surely was. Not everyone would have done that.'

I was annoyed to feel myself beginning to blush.

'I thank you, anyway,' Roland continued. He stood up and walked towards me. He was two or three inches shorter, dark-haired with large, open grey eyes, an almost pallid complexion and fine features verging on the delicate – a master-sculptor's version of his father, in fact. His look contained something transparently naïve, even ethereal and yet I sensed the immediate contradiction in the way he held my gaze, with its suggestion that somewhere within him lay his mother's resolve. He held out his hand with a formality to which I was quite unaccustomed. I shook it tentatively and was surprised by the firmness of his grasp. He reached for his hat, which lay on the table, and made for the door. 'We should go and see what my father has planned for us.' He made no attempt to conceal his eagerness.

'Be sure and be back by midday.' His mother glanced at me as I followed him out. 'There will be food for you both.'

Everyone else in the village went to work with bread and cheese in their pockets. A midday meal at home was unheard of. Maybe this was something to do with Roland's condition. Or maybe it was just that for the time being I was to remain under scrutiny. I nodded to her and we went downstairs.

'Now,' said the steward, blowing on the tips of his mittened fingers, 'a task for willing hands . . . It has been on my mind for some time that our lord may wish to reinstate the deer park – when the present troubles are past, that is. The brushwood should be cleared before the palisade can be rebuilt. There are billhooks and axes, newly sharpened, over there.' He pointed to a corner of the storeroom, then fumbled beneath his desk and produced a flint. 'Take care not to burn down the trees.' I forebore from telling him that I now had a flint of my own.

With implements over our shoulders we set off for the deer park, a semi-circular enclosure of three or four acres that stood between the rear of the manor and the forest. Over the years it had fallen into a dreadful state of disrepair. The palisade had rotted and collapsed early on and the deer had escaped back into the forest which, in turn, had begun to encroach on the untended parkland. Its furthest reaches had now reverted to a wilderness of thicket and saplings whose clearance, in my opinion, represented a month's labour for a dozen woodsmen.

Although I had no cavil with hard work – I was perfectly used to it – our orders struck me as quite unwarranted since our lord, on his rare appearances, confined his attention almost exclusively to the contents of the salt-store and evinced no interest whatever in the improvement of his surroundings. It took me a moment or two to realise that our task had an altogether different purpose – that of getting Roland out of the house and restoring some colour to his almost ghostly flesh, while at the same time keeping him as far from the village and the villagers as possible, for as long as possible. Whether we cleared the place, or even made an impression on it, was unlikely to be of any consequence.

We tramped across the sunlit park with our breath clouding before us in the still, cold air. A partridge creaked somewhere in the neighbouring pasture, Roland's footsteps crunched in the snow behind me and it occurred to me that the last time I had been accompanied by the tread of younger feet, they had signalled a companionship which was as familiar as the feel of my own skin. I glanced back at him and wished that he was not with me.

He appeared to be concentrating deeply, but he sensed my attention and looked up.

'Creb.' He said it almost to himself, sounding puzzled.

I waited.

'It's a curious name.'

'Does all right for me.' I looked away and continued walking.

'Is it your birthname?'

I shook my head.

He persisted. 'What is your proper name, then?'

I felt disinclined to answer. It was, in fact, one of those garbled diminutives which so often arise when a younger sibling has been unable to master the whole mouthful.

There was a pause, then: 'It wouldn't be Cerdic, would it?'

I shook my head again and quickened my pace.

'Hmm . . . then what could it be? . . . unless it has nothing to do with your real name at all . . .' He was beside me now, struggling to keep abreast.

I turned to him and glared. 'It's not your business.'

I was expecting him to change the subject, or fall silent, or walk on ahead, but his reaction took me quite by surprise. He tugged me gently to a halt and said: 'I apologise, Creb, I didn't intend to pry. It's just that . . . there's more to everything than you first see or hear . . . and, well . . . names are important . . . they tell you things. I believe they do, anyway.' There was something entirely ingenuous in his look, a kind of honest pleading which made it almost impossible to feel the resentment I wished for.

I was startled to hear myself saying: 'It's Christopher.'

Roland looked at me, then nodded to himself: 'The Christ-carrier.'

Where my parents had come by it, I do not know. How they, of all people, had then bestowed it on their son, I know even less. It was not at all common and I remember even at an early age being very much discomfited by it and feeling greatly relieved when it came to be superceded by Creb – which was even stranger in its way, but in sound, at least, a good deal closer to the grunted monosyllables that passed for names amongst most of the rest of the villagers. I had, of course, been told the story of the broad shoulders, the stout staff and the crossing of the swollen river – and had perhaps felt a little throb of childish pride – but I had long ago forgotten about it and was startled to find that someone else was familiar even with the name, let alone its significance.

'But you prefer to be known as Creb?'

'Yes.'

'Naturally. You're accustomed to it.' He looked at me closely again and nodded. 'Yes, it has a good sound to it – not too hard, not too soft. It becomes you.'

I was not used to being spoken to so personally, so directly and I did not quite know where to look.

'How do you know,' I asked, 'about the name?'

He thought for a moment. 'From our lord's chaplain, most likely. When I was a child I used to play with our lord's son – when he was here. We were about the same age. We were a good deal of time in the hall, with all the fancy people. I believe the chaplain took a liking to me. He told me all manner of things.'

'Oh.'

'Yes, remarkable things. For instance . . . at a royal banquet they ate three hundred herons, two hundred swans and a hundred and twenty peacocks – just to whet their appetites. Imagine it, Creb – thirty-one score of birds, large birds, before they even caught a whiff of the beef and venison! And they could see the smoke from the kitchens a hundred miles away. That's what he said. I don't believe you could see smoke from anything for a hundred miles but it must have been a feast to remember, all the same.'

I had never heard anything like this before in my life. My mouth was hanging open and I shut it quickly, not wishing to appear too impressed. But Roland's look had become cheerfully absent, as if he were rummaging his memory for another tasty morsel.

'Do you know of oliphaunts?' he asked.

The word was dimly familiar. I shrugged.

'Oliphaunts, Creb, are the strangest creatures in the world. They live in a hot country, far away. They're as large as – well – your cottage, I'd say. They have immensely long noses. So long that they brush the ground when they walk. And through these noses of theirs they can blow water, like a fountain. To keep themselves cool in all the heat, you see. It's true. There was a knight who came to the manor once. He had seen one. Think of it!'

His enthusiasm for what he was telling me was so evident that I found myself admitting that perhaps he did not mean to impress me at all. On the contrary, by sharing what he knew, he appeared to increase his own delight in it.

'What else do you know?' I asked.

Roland closed his eyes and thought, then: 'How about this.' He lowered his voice conspiratorially. 'There are learned men who can turn base metal into gold.'

I looked at him blankly.

'They take a lump of lead and heat it in a crucible. Add a pinch of something to it and – lo and behold – it transmutes. Turns into gold. Alchemists, they're called. Very learned indeed – and very rich too!'

'Sounds like witchcraft.'

He shook his head emphatically. 'Oh no. Far from it. It's a kind of philosophy – but with matter. The highest form of learning there is.'

I was beginning to feel a little in awe of him. I knew nothing, literally nothing of the world beyond the forest and the estuary. Even the geese on the marshes knew more than I did. Yet here was this sickly-looking youth, three years my junior, whose mind appeared to be stuffed to bursting with extraordinary knowledge.

'Do you know book-learning?' I asked tentatively.

'Some. My father has our lord's psalter while he's away and he keeps account of the rents and disbursements and stores and so on. Scarcely interesting. But it keeps me in practice.

If he could read, then perhaps also he would be able to decipher my little drawing – if I allowed myself to trust him. But I was not ready to do that yet, not by a long mark.

'Wonder what happened to the wheelwright . . .' I mused, deliberately changing the subject.

'They say he fled into the forest.' He looked up. 'Have you ever been there, Creb?'

'No,' I replied, rather too hastily. 'Well . . . yes . . . but only to the timber-wood and the pannage. The wheelwright, he must have been mighty scared.'

'Yes,' said Roland with sympathy, 'I imagine he was.'

'I reckon the reeve made it all up.'

'My father mistrusts the reeve.'

'People say he's out for revenge now.'

'On what account?'

'Being called a liar.'

'By the wheelwright? But he's gone.'

'Not the wheelwright. There was someone else called him a liar . . .'

He shook his head slowly. 'That must have been after I had my fit.'

So he really did not remember. The malicious urge was almost irresistible . . .

He looked at me uncertainly. 'It was . . . nothing to do with me, was it?'

I forced a laugh. 'No. Someone near the front. I never saw who. Just wondering how much you remembered.'

'Not very much,' he replied softly. 'Shall we talk about something else?'

The rest of the morning passed without event. After a desultory inspection of the undergrowth, we rolled up our sleeves and began hacking away at bushes and brambles.

It soon became apparent that Roland's deftness of mind was not echoed in limb. He was disastrously lacking in co-ordination and seemed incapable of striking what he aimed for, or even controlling the force of his stroke. After he had missed amputating his foot by a whisker, I took the axe from him and relegated him to the removal of the felled brushwood into piles for burning. It was gratifying to discover that there was something at which I was his better but he accepted his demotion with such good humour that I knew it would be churlish to press the advantage.

The morning wore on and I found myself becoming increasingly fascinated by him. He seemed able to swing from exuberance to reflection with an almost musical fluency: voluble and energetic at one moment as he cheered on my axe-strokes, then helped me heave the toppled sapling into the open; serious and silent the next as he examined an animal track or listened intently to my halting explanation of some phenomenon that interested him – for it transpired that I also knew more about nature than he did.

As we made our way back to the manor at midday, I realised that most of the villagers I knew were as interesting as clods of earth by comparison and, as we entered the steward's quarters, I found myself wondering, more than a little anxiously, whether that was the way I appeared to him.

I sat down at the table feeling suddenly self-conscious. The evidence of status, which I had been too preoccupied to notice during my interview with the steward was all around me: the stone hearth, a couple of iron-bound oak chests, candles in pewter sticks, heavy wall drapes and a faded tapestry – our lord's hand-me-downs more than likely, but sumptuous nonetheless in comparison with the crude furnishings of our smoke-darkened cottage – and there, miraculously, a glazed and leaded window in the recess, flooding the far end of the room with sunlight now that the curtains were pulled back. It must have been the only one in the village, for our lord took the manor windows away with him in a cart each time he left, frames and all.

We were joined by the steward, his wife and his widowed sister, a small silent woman with a hare lip. She had come to visit shortly before we received news of the pestilence, so Roland had explained, and was now pining for her new suitor, a fishmonger in some distant town. She was fond of her brother, he had added, but hated the country and everything to do with it, particularly now that she could not escape it.

Throughout the meal, Roland prattled cheerfully about the morning's events, referring to me frequently in tones which were a little too full of admiration for my liking. 'Creb did this . . .' 'Creb told me that . . .' If anything, I thought, it should be the other way around. But it was clear that the double novelty of a morning away from his family and the companionship of someone closer in age had inspired him beyond measure.

Still somewhat awed by my surroundings and unused to such lively conversation at table, I ate in silence and nodded from time to time as Roland chattered away. But within a short while the luxury of a hot vegetable broth, the heat from the fire and his family's apparent approval began to dispel my sense of inferiority and a warm glow of contentment spread through me. I glanced at the animated figure opposite, eyes bright and cheeks flushed, and winced to think how near I had come to devastating him. His company that morning had afforded me more pleasure than perhaps I cared to admit. And in the same breath I understood that here were echoes, albeit faint ones, of the affinity I had felt for my brother. I would have to be on my guard against that from now on.

We finished our meal, wrapped up against the cold once more

and set off into the deer park. We had gone only a little way when Roland stopped and looked at me.

'You're quiet, Creb.'

'Am I?' I endeavoured to sound nonchalant but I was shocked at the speed with which he seemed to have sensed my change of mood.

He was pensive for a moment. Then, with great sympathy, he said: 'Your brother. I know.'

The horror on my face must have been quite apparent for he put his hand to his mouth and said immediately: 'Oh, Creb, I am sorry. Truly sorry! I say such foolish things.'

I hardly heard him, for all at once I wanted to run off into the forest and scream . . . to hurl myself into the snow and rage at my father who would not talk . . . at my mother who talked with her eyes . . . at our neighbours with their discreet condolence . . . at the steward who had tricked me into this . . . and I wanted, most of all, to be far away from this uncanny youth who made me feel suddenly so exposed, so naked.

I turned to flee, lost my footing and fell heavily. I lay still for a while, partly winded. Then, as my breath returned, I sat up and found myself beginning to weep again.

Roland hovered awkwardly at a distance, a stricken look on his face. For some moments I struggled to bring myself under control, then rubbed a handful of snow over my face and got to my feet.

'How did you know?' I asked, walking towards him. He looked at me with hesitation, then replied gently: 'Creb – everyone knows. And you seemed so distant just then, so troubled . . .'

'You guessed?'

'Of course.' He arched his brow a little, then smiled. 'How else could I have known? I can't look inside your mind . . .'

'No,' I said, feeling suddenly idiotic and weak. 'We'd best get back to work.'

We walked on in silence and gradually I became aware that there was something in Roland's presence that was starting to draw me from myself, like the action of a poultice on a wound. I resisted it for a while, then found myself saying: 'Shall I tell you what hapened?'

'If you wish.'

So I told him, casting back for every detail, however small and insignificant. Roland listened seriously and attentively, interrupting only once to enquire about the thickness of the ice. We halted by our morning's work and continued to stand there as I reached the end of my story. A kestrel hovered in the sunlight above our heads.

Roland remained silent for some while after I had finished. Then he looked at me and I noticed for the first time the strange maturity, the suggestion even of wisdom, that lent such a grave expression to the boyish face.

He pointed into the sky. 'What food does that hawk find?'

'Mice, voles, shrews . . .' I replied. 'It's a kestrel.'

'Then in there,' he said, raising an arm to the undergrowth behind us, 'is a mouse or a vole or a shrew. That kestrel knows its name.' He paused. 'And something, somewhere, knew your brother's name, Creb. It was his time.'

I glanced up at the hawk with its wings spread and tail fanned against the winter blue. His time to die . . . as it had been my time to go the forest and not to die . . . to find the monk and the drawing . . . to put the knife in Roland's mouth and end up here . . . his companion . . .?

At length, the kestrel abandoned its quarry and dipped away towards the forest. Roland watched it and nodded thoughtfully. 'Well, perhaps not this time. But there will be another one, in some other place.'

'Yes,' I said, curiously grateful for this small glimpse of certainty.

Five

hree days later, the cold weather broke. A thaw set in overnight and by mid-morning low cloud had crept up the river, bringing with it a persistent drizzle that washed colour and definition from the landscape. All around us the undergrowth dripped continuously, while the forest, weighed down by cloud, seemed slowly to be wringing itself out. In the open parts of the deer park and the pastures beyond, the pristine white mantle was swiftly becoming stained a dirty, sodden grey.

Roland and I worked on, but despite his constant feeding of the large fire which consumed brushwood as fast as I could cut it, he clearly felt the chilling dampness a good deal more than I did. I watched his movements becoming lethargic, his expression increasingly dejected. Although he did not complain, I was reminded that my charge was no weather-hardened village lad who would continue working with bovine resignation until his hands were so cold he could no longer hold his implement, the mud so thick he could no longer lift his feet.

My charge. No, my companion. My friend? If a shared curiosity constituted friendship, then we were already friends for we had plundered one another relentlessly – I for the nuggets in Roland's treasure house, he for the nuts in my squirrel's hoard. If at first I had wondered of what possible value my stolid company could be to this quick-witted, mercurial lad, I now realised that he was as intrigued by my simple knowledge of nature and the land as I was dazzled by his tales of exotic places and events.

But if this was indeed friendship, then there was much about it that was new to me. My one previous friend had been my brother and I had been lucky to have loved him almost without reservation:

his friendship I had taken as much for granted as his kinship. Now I was becoming slowly aware that friendship was favoured as much by chance as by the intrinsicalities of blood; that differences in circumstance were no impediment to it and that it was advanced on the flow of an inexpressible current. I sensed dimly also that there were complexities, that the current would not always run steady and true, but for the time being the course demanded no more than a little gentle probing here, a little testing there. So now, as Roland struggled back and forth to the fire with armfuls of brushwood, I watched but said nothing.

At midday I laid down the axe and put on my jerkin and saw the relief on his face as he threw a last branch onto the blaze. We walked back in silence through the melting snow. Eventually he looked at me and said, without accusation: 'We should do something else this afternoon.'

I nodded.

For a moment he looked pensive. 'You've been quiet . . . thinking . . . puzzled . . .'

The little drawing had been in my mind all morning, as clearly as if I held it in front of me.

'How do you know such things?'

'From your face.' He gave a little smile, as if it were the most obvious thing in the world. 'I've told you before – you can learn from faces.'

'Hmm.'

'A secret?'

I deliberated – for a second too long – before nodding.

Something mischievous crept into his eye. 'I have one too.' He paused. 'Mine for yours?'

'Maybe.'

'But maybe not . . .?' He was grinning now, the grey eyes chiselling away at my resistance.

'I didn't say that.'

'Well?'

'I'd need to go home . . .'

'So . . .' He tipped his head towards the village.

I left him guessing until we were almost at the manor, then turned away for the lane.

'When we've eaten,' Roland called after me. 'Mine first, then yours.'

I returned to the cottage and fetched the parchment from beneath my mattress where I had kept it since the wheelwright's trial. As I walked back to the manor something hard and cold struck me behind the ear. I turned round to see the reeve's eldest son standing in the lane behind me, surrounded by his cronies. His reputation for trouble-making was almost equal to that of his father and I had always taken care to avoid him. This seemingly unwarranted assault was therefore all the more startling. He raised his hand and yelled something abusive, then let fly again. There was a hail of snowballs as the others took up the refrain: 'Witch-lover!'

I turned my back to them and walked on, my ear stinging.

I was beginning to enjoy the company of the steward and his family. There was a warmth in the attic chamber, a reassuring sense of mutual dependence and support, an openness of affection. The latter made me feel a little shy but I also found it touching – more so still as I came to realise that, quite unobtrusively, it was being extended to me, too.

Today, as on most days recently, the conversation turned on the steward's concerns. All the hearsay, all the opinions of his fitness or otherwise for his position, were gradually becoming illuminated as I listened to him wrestling aloud with the problems brought about by the pestilence, the shortage of provisions, the growing dissent in the village. It seemed clear to me that he cared too much, felt too responsible and, while I could not help admiring him for it, I was also beginning to understand that a firmer hand than his would have already brought the situation under control without any permanent loss of sympathy or respect. The reeve, needless to say, was the principal fly in the steward's ointment – all our ointment, come to that, although most would not have admitted it – and it appeared that he had already made his opening move. For several days now a rumour had been circulating to the effect that there were more provisions to be had than were actually being dispensed, and that while everyone else was slipping rapidly towards bare subsistence, we at the manor continued to live high on the hog. Whether or not he was at the source of the rumour, the reeve had been swift to make capital of it, letting it be known that he was seeking volunteers for a deputation which would shortly wish to inspect the storerooms. This in itself caused the steward no great alarm since his conscience was clear, but it was evident that the threat held broader implications that worried him deeply. From the way he spoke there was also a suggestion that the reeve might be employing other tactics he did not care to discuss before us and as I realised this, my thoughts turned at once to what had happened in the lane, on the way back to the manor. What worried the steward most deeply, however, and about which he made no pretence, was that he did not know where it was all leading. Again, I found myself thinking that a more determined man would have called the reeve to account at the wheelwright's trial. Given the fickleness of the audience-cum-jury, he would undoubtedly have carried the day. But that moment had passed.

'Someone will have to go out soon,' he said, taking a sup of his broth. I had begun to notice in recent days that it was tasting increasingly watery.

'How much longer do we have?' asked his wife.

The steward frowned. 'A week, ten days at the most.'

'Then what will happen?' The steward's sister seldom spoke. Now she sounded anxious.

'We will start to go hungry.'

'But do you not grow what you need, you country folk?'

The steward shook his head sadly and pointed at the window. 'Not on this poor land. Were it not for the salt there would be no manor here, no village, no animals. As it is, our lord makes a small allowance from the salt revenues to tide us through the winter. All very well when we can go to market . . .'

'If it was safe for us to go out,' said his wife thoughtfully, 'the pedlars would surely have returned by now.'

'If there are any pedlars . . .' He ran a hand through his hair.

'Our lord said he would send word when it was safe again,' Roland observed.

'He did,' said his mother with an ironic smile, 'but – even if the pestilence has passed, the countryside may not be safe for a messenger. Your father is right. Someone should go and see . . .'

'Particularly with the reeve in his present humour,' added the steward gloomily. 'I do not wish to give him the slightest opportunity to say that we are not doing all we can.' He scratched his chin. 'Perhaps he might care to go himself?'

We laughed, but with little conviction.

'Do you go hungry at home, Creb?' the steward's wife asked.

'We've enough,' I said non-committally. The truth was we had finished our one small side of bacon some time ago and the meagre weekly allowance of flour and dried vegetables did not go very far. Were it not for the chickens we would already have had rumbling bellies and not everyone had chickens.

She gave me a gently questioning look but did not press me. Roland left the table and went to the window.

'It's still raining,' he announced. 'Too wet for the deer park, I think.' He coughed ostentatiously but his father did not seem to have heard or noticed.

'It's raining,' said his wife, touching his arm. 'It would be better if Roland were not outside.'

'Very well,' replied the steward, focusing briefly, 'but he should do something useful.'

'There's the hall,' said Roland quickly. 'It could do with new rushes, what with all those feet the other day.'

A more pointless exercise I could scarcely imagine. Before the trial, the place had not been used in months and it was unlikely to be used again for months to come. But the steward was clearly glad of any suggestion that relieved him of yet another decision.

'The hall,' he said and nodded, his gaze wandering into the middle distance again.

Roland walked back towards the table. Passing behind his aunt's chair he glanced pointedly at the back of her head and gave a surreptitious wink whose significance escaped me.

'Do you have it?' he asked, as we made our way across the courtyard to the hall.

I nodded, patting my jerkin.

Roland grinned.

'Well . . .?' It was beginning to seem somewhat puerile – this business of secrets – and now that I had made up my mind, the parchment was burning a hole in my shirt.

'In a little while. You'll see.' He bent down and gathered an armful of mildewed rushes from the floor.

It took us some time to clear the old stuff from the hall. When we had finished, we passed through into the buttery and threaded our way between the wine casks to the small store at the rear. Bundles of aromatic herbs hung from the ceiling and sheaves of rushes were stacked on lengths of timber which allowed the air to circulate beneath them. I could hear movement in the kitchen which lay beyond and the smell of woodsmoke mingled with the tang of herbs.

'Is someone cooking?' I asked.

Roland shook his head knowingly.

'What, then? Who uses the kitchen when the manor's empty?'

'Aha!'

I told him that his smugness was irritating me but he merely reached for a bundle of herbs and gave me the particular smile – head to one side and eyes impishly wide – which he had previously employed to great effect when I was in the mood to chide him.

'Just a little longer. Patience, Creb!'

I followed him back to the hall and we sprinkled herbs over the floor. When we were done, Roland stood back looking exceedingly pleased with himself.

'We'll lay the rushes later,' he said. 'Come.'

He returned to the buttery where he put his finger to his lips and climbed onto a cask, indicating that I should do likewise upon its neighbour. I clambered up to find my head level with a ragged gap of roughly a handspan between the top of the wall and the thatch. Whether the wall had crumbled through neglect or been knocked out for ventilation, the resulting space provided a clear view through to the kitchen which was now hazy with smoke from the fire and steam from the cauldron hanging above it.

Roland winked broadly, then pressed his face to the gap as we heard the sound of something heavy being dragged across the floor. In a moment, from the far end of the kitchen appeared his aunt, heaving a large wooden tub towards the fire. She was flushed with the effort. Wisps of mousy hair had escaped from her headdress and clung to her forehead. Once she had positioned the tub in front of the fire she fetched a jug and began to fill it with hot water from the cauldron. As she went back and forth to the tub I squinted around the room for a pile of washing but could see none.

Now she rolled back her sleeve and dipped her wrist into the water. She gave a little exclamation, withdrew it hasily and disappeared from view. After several trips with the jug it appeared that she finally had the water to her satisfaction. She produced a small phial, unstoppered it and let a couple of drops into the water and a thick scent of lilies immediately wafted up on the steam. Then she stood back and began to remove her clothes with precise, practised movements, folding each item methodically and placing it on a stool before the fire – quite unnecessarily, I thought, if she was intending to wash them. At length she stood only in her shift. Roland nudged me and winked again as she shivered briefly, then pulled the shift over her head and I realised that it was not her clothes she was going to wash, but herself. I looked on in astonishment as she stepped into the tub and sank down into the steaming water with a sigh of pleasure.

For some time she simply sat there, her head resting on the rim of the tub, a look of mild contentment on her upturned face. I glanced at Roland with his feet braced on the cask, his nose pressed to the wall. He turned to me and grinned and I found myself thinking what a very curious secret this was. For although to witness someone taking a bath was certainly a novelty to me, Roland had clearly seen it several times before and yet he still appeared to be enjoying himself hugely. I could understand that his aunt might not be overjoyed at the notion of an audience and that she should therefore be kept in ignorance of his attention, but did he really find pleasure, that clandestine delight which is the essence of a true secret, in watching a naked and nondescript middle-aged woman half submerged in hot water? Evidently he did. It shocked me a little to realise that, sophisticated as he was in other things, where women were concerned he was still gratifying the adolescent urge to peep and pry.

'She'll start to wash in a moment,' he whispered.

Now she fumbled in the water and produced a square of sodden cloth which she ran gently across her face, her neck and her shoulders. A transformation had come over the normally shuttered features: she was half smiling to herself and her eyes had grown moist and dreamy. I wondered whether she was thinking of the fishmonger. After a while she eased herself up and, as her shoulders came clear of the water, her breasts were partially revealed. A brief glimpse as she had stepped into the tub had shown them to be small and flaccid but, buoyed up by the water, they now quivered gently as she rubbed them with the cloth. It must have been a pleasurable sensation for she began to rock her head from side to side and hum softly. There was a soft hissing and I turned to see Roland puce in the face as he held his breath in his attempt not to laugh. For a few moments he contained himself, then the effort proved too much and he let slip a spluttering cough. His aunt turned sharply and glanced around as both our heads ducked below the wall. There was a commotion at the far end of the kitchen as the door was flung open to the sound of raised voices. Something squealed loudly and was echoed by a horrified shriek from Roland's aunt. We lifted our heads cautiously to see her standing up in the water with a look of abject horror as she attempted to cover herself with her hands and at the same time wave away two large pigs which were advancing inquisitively towards the tub.

When they were a few feet away they halted and peered myopically at the naked figure before them, wrinkling their snouts in an attempt to identify the unaccustomed fragrance. Roland's aunt now stood stock still in the centre of the tub with one arm clasped at her breast and the other hand spread at her groin. Her marble-white flesh was stippled with goose-pimples.

There was a peal of laughter from beyond as a turnip sailed through the air and struck one of the pigs on its hindquarters. It squealed indignantly and began to caper crazily, round and round. Unperturbed, the other trotted forward and lifted its snout to the rim of the tub. Roland's aunt gave another shriek, stepped away as far as she could, lost her balance and toppled backwards out of the tub, drenching the pig which grunted furiously, shook itself and made straight for the stool to console itself with her gown. Roland's aunt did not get up.

Roland stopped laughing. 'I think . . . perhaps . . . we should go and . . .'

'We should,' I said curtly.

We left the buttery and made our way quickly to the kitchen. The door was open and we paused on the threshold.

'Aunt . . .?' called Roland, hesitantly.

There was no reply so we stepped inside. She had drawn the stool up to the fire and was sitting, still naked, with her back to the door and her head in her hands. One of the pigs had disappeared. The other was lying in a puddle of water by the tub, chewing ruminatively at her gown. She did not look round as we came in.

Roland gave me an uncertain glance, then walked slowly towards her, picking up her shift from the floor as he went.

'Put this on,' he said quietly. 'You'll catch cold.'

She reached out a hand without looking up and took it. Then

she half-turned to wave us away and I saw that her face was wet with tears.

Roland retrieved her gown, laid it on the stool and shooed the pig out of the kitchen as we left. We walked back to the hall in silence and began to lay the new rushes. For some time we worked without speaking, deliberately keeping our distance, avoiding one another's gaze. It was not until we were nearly done that Roland glanced across to me and said: 'You're angry with me . . .'

I took my time to finish scattering my bundle of rushes before replying: 'I am. And ashamed.'

'She didn't know . . .'

'Makes no difference.'

'But we weren't doing any harm, Creb.'

'Mother of God, Roland! Spying on a woman in her bath – what kind of secret is that? Well? A clever lad like you – educated and all? It's . . . it's . . . beneath you!'

His eyes flashed briefly, then his shoulders slumped and his gaze fell to the floor. He had become very pale and it occurred to me that I might have overstepped the mark. The strange thing was that I had felt more disappointment than shame or anger. And yet, what right had I to feel any of these things? For the mere three years' difference in our ages, I was to all intents and purposes a man whilst he was yet a boy, and in some respects the threshold of maturity still yawned between us . . .

He shrugged disconsolately and fiddled with a rush stalk. 'I apologise, Creb. You must find me a trial at times.' He sighed and then, as if having read my thoughts, continued: 'I know I'm . . . inexperienced in many things. Women, for example.' He looked up at me now. 'They interest me. It's natural, I suppose, but I know nothing of them. Never even really seen one . . . until Aunt came along. You know all about women, don't you, Creb?'

'Mmm . . .' The naïvety, the honesty, were so utterly disarming. 'I was wrong to shout. I'm sorry. Who was it let the pigs in?'

'Youngsters from the village.' He sounded grateful. 'They've plagued her since she first came here – for being from the town. That's why she doesn't go out any more. She hates this place.'

It occurred to me that the reeve's son might have played some part in this. Sparing him the detail, I related what had happened on the way to the manor.

He nodded. 'He's a serpent, the reeve's son. He gets the others to do what he says – like his father. That's why Father sent us to the deer park.' He caught my look of surprise and smiled. 'I may be inexperienced, Creb, but I'm not a fool – it's because of my fits, I know. And now that we're going to go hungry, there'll be further trouble – for all of us.' He paused, looking at me solemnly. 'You . . . you don't have to stay with me, you know. My father would let you go.'

I shrugged. 'Rather you than them.'

He held my eye and nodded slowly. 'You mean that, don't you?' 'I do.'

A slight flush came to his pale cheeks. 'I'm glad. Thank you.' He brightened, the mischief returning. 'Then there's still the matter of . . .'

'I know.' I reached inside my jerkin but at that moment the door swung open and the steward came into the hall, looking more than usually concerned.

'Do you know what has happened to your aunt?' he asked without preamble. 'She has had a fearful shock and refuses to speak.'

'Someone let pigs into the kitchen while she was taking her bath,' Roland replied.

The steward shook his head sadly. 'Do you know who?'

'We didn't see. But Creb thinks it may have been the reeve's son.' He glanced at me for confirmation and I nodded.

'That would not surprise me,' said the steward, scratching his head. 'What am I to do, then?' He stared into the empty hearth, then turned abruptly and strode to the door. 'You had best finish this quickly and come inside – both of you.'

Roland grimaced with disappointment as his father left. 'Is there time now?'

I shook my head. 'It'll keep till tomorrow.'

Behind the hall ran a passage where we had stacked the bundles of old rushes. The door led out to a patch of rough ground at the rear of the building where a slight mound beneath the melting snow betrayed the presence of a burning-place. Since it had now stopped raining, we began to take the rushes outside. On the third or fourth trip I emerged to find Roland sitting on the ground, his eyes closed in pain. With one hand he was rubbing his cheek. In the other he held the remains of a hard, wet snowball in the centre of which was a stone the size of a blackbird's egg.

I glanced around and could see no one, but beyond the end of the kitchen, some twenty paces off, was a hedge quite thick enough to conceal the thrower. I went towards Roland and another missile struck the rushes at my feet. He was paler than ever and struggling to maintain his composure. I helped him to his feet and began to steer him towards the door but as we approached, it swung to and I heard the key turn in the lock. Roland glanced at me in alarm as we set off hurriedly along the back of the hall, away from our invisible assailant. We had gone only a few paces when three youths appeared round the corner at the far end. I stopped and looked over my shoulder to see the reeve's son and two other lads emerge from between the kitchen and the hedge. Escape lay out across the deer park towards the forest but I felt suddenly angry and ready to bloody a nose or two for both our sakes.

We stood our ground as the youths advanced from both sides, until they were close enough to encircle us. The reeve's son stepped forward and spat a gobbet of phlegm at Roland's feet. It left a neat hole in the wet snow.

'Went to the rescue, did you?'

Roland said nothing but I noticed his pallor tinged with a faint flush. His breathing seemed to be quickening and I realised, to my regret, that I should do whatever I could to avoid a fight.

'Saw 'er dugs, I expect,' said one of the others, smirking. 'Like 'ounds' ears.' There was laughter.

'If she don't like pigs she shouldn't be 'ere,' said another. 'Ain't no place for townfolk.'

'No,' said the reeve's son, 'no place. And,' he took another pace forward and stared at Roland, 'no place for witches, neither.'

Roland's eyes widened. 'Wha . . . what do you mean?' He was beginning to pant softly.

'What I mean,' said the reeve's son with slow menace, 'is know-alls – little cock-fiddlers – who think they can tell what other people have been about, when they weren't there to see with their own eyes.'

Roland glanced at me in bewilderment. He seemed to have developed a tic in both eyelids. I stepped closer to him and the reeve's son glared at me.

'You stay out o' this, witch-lover.' He turned to Roland again. 'You know what's on my mind, brown-breeks. Calling someone a liar – in front of everyone.'

'I . . . I don't understand,' said Roland. He was beginning to sway. I put an arm around his shoulders and said, with all the restraint I could muster: 'Let him be.'

One of the youths stuck his face close to Roland's. 'Don't look too good.' He contorted his features grotesquely. 'P'raps he got the pox – or the pestilence!'

'Not him!' exclaimed another. 'Wouldn't dare leave the manor.'

'Well, now it's time he did,' said the reeve's son. He shook his fist in Roland's face and I could no longer contain myself. I felt no pain as my knuckles met his teeth. He staggered back, spitting blood, then lunged at me as another youth kicked me hard on the shins and I felt my arms grabbed from behind. A fist caught me in the stomach. My lungs seized, my knees buckled and I went down choking for wind. I felt a foot in my ribs, another against my shoulder and as I struggled to cover my head I dimly heard a curious gargling sound. The blows suddenly stopped. For a moment there was silence and I lifted my head to see my assailants staring open-mouthed at the ground behind me. Then they turned and fled without another word.

I rose painfully to my feet and looked around to see Roland lying on his side, his knees drawn up and his body racked with convulsive shudders. Through the half-closed, fluttering lids, I could see only the whites of his eyes. His lips were flecked and he was breathing very rapidly, like an exhausted dog. I knelt down and took his head in my hands. It was harder, this time, to suppress the sense of revulsion at a body whose inhabitant had so manifestly lost control; it reminded me of a large, pale, wounded insect. But at the same time it clamoured for my protection and I pulled him close against me and willed the fit to pass.

In due course it did. His breathing returned to normal and I felt his limbs unlock and his body settle loosely on the ground. I said his name a couple of times but he did not respond, so I lifted him up and made my way, pausing frequently, to the storeroom, where I laid him at the foot of the stairs.

I was almost at the door when it opened and the steward emerged, his face drawn with tension. Glancing past me to the limp figure below, his eyebrows rose in horror.

'He's all right,' I said hastily. 'It's passed.'

He gave a little grunt of relief and hurried down the stairs.

Together we carried Roland into the sleeping-quarters where we laid him on his bed. The steward ran his hand across his son's brow, then left his wife to take charge and led me back into the main chamber.

'What happened?' he asked.

'We were set on,' I replied. 'The reeve's son and some other lads.'

The steward's eyes hardened. 'Was it, by God! I was on my way to see his father anyway.' He hurried back into the sleeping-quarters where I heard him conferring softly with his wife. Then he returned, strode to the door and went loudly downstairs.

A moment later his wife came out. She thanked me for helping Roland, then said quietly: 'I think you should leave now. You have done all you can.'

I turned to go.

'And, Creb . . .'

'Yes?'

'You mean so much to him.'

Six

he light was failing as I left the manor and made my way down the lane. Ahead, the elms loomed naked into the dusk, dwarfing the cottages beyond whose squat forms were silhouetted like so many loaves against the dull glint of the river. There was an unnatural stillness, broken only by the soft patter of drizzle in the hedgerows and it took me some moments to realise that at an hour when folk would normally have been returning from their work, there was not a soul to be seen. Although I did not understand why, my first reaction was one of relief for it meant that I was spared the possibility of further confrontation. But as I walked on I became increasingly uneasy.

I reached our cottage and stopped on the threshold. What little furniture we had was smashed beyond recognition. The partition between the living- and sleeping-quarters had been torn down and grey light gaped through holes in the thatch. Shards of earthenware littered the floor. A cat hung impaled on a pitchfork protruding from one of the wall timbers. Blood dripped from its muzzle.

In the gloom and the chaos I could see a figure sprawled awkwardly against the wall in the far corner of the room. It was my father. He had a deep gash at his temple from which blood had flowed copiously over his face but the sound of his breathing told me he was still alive. There was another sound, a soft whimpering, coming from what remained of the sleeping-quarters. I picked my way through the debris, calling out for my mother and saw a movement amongst the bedding and straw and splinters of pallet strewn across the floor. She had burrowed under the mattress and lay there quivering with distress. I helped her out and half-carried her back into the living-quarters where she sank to the floor and began to sob inconsolably onto my shoulder. It shocked me profoundly to see her in such a condition and not knowing what to do or say I simply held her as tight as I could and stroked her head. I found myself staring past her at the fragments of a little wooden figure my father had been carving – a knight with lance and shield – and wondering inconsequentially whether it could be put back together again.

'Oh, Creb . . . they came in here like . . . like wild heathens . . . staves and axes and forks they had . . . calling us dirty, dirty names . . . shouting and laughing . . .' Her voice was muffled and tremulous as she tried to control her sobbing. 'See . . . I always thought . . . one good deed . . . begets another . . . least, till now . . . You did . . . doing . . . a good thing for that lad . . . and this . . .' she tried to gesture at the devastation but her arm fell limp at her side '. . . this is what you get.' She began to weep again.

My father gave a sudden groan, sat up and looked across at us with eyes quite blank. My mother turned towards him and shook her head tearfully; a small, hopeless gesture.

'No helping him now . . . he didn't even try and . . . stop them . . .'

This visible collapse of my parents was unnerving me. I felt bewildered, gripped by a sudden childish fear, unable to order my thoughts. 'What should we do?'

'You should go where you went before . . . to the forest . . . wherever it was . . .'

'But you – and Father . . .'

She shook her head and shuddered. 'It's you they want, not us . . . anyway, they can't do worse by us now . . . go . . . take the lad with you . . . take them all if you can . . .' She sank back, her shoulders drooping with exhaustion.

I hovered indecisively.

'For the love of God, son – go!'

I stood up and wrenched the pitchfork from the wall, dislodging the corpse from its tip. Then, stooping to kiss my mother, I left the cottage and walked briskly away in the dusk. I noticed that most of the other cottages were still unlit and tightened my grip on the smooth wooden shaft of my weapon.

I was close to the well when I heard the dull clamour of voices at the other end of the village. Even at this distance there was a discernible tension to the sound: a rumble swelling rapidly to a taut crescendo, dying away, then swelling again. I glanced in that direction and saw the flicker of torchlight amongst the silhouetted cottage roofs. I began to run.

Beyond the cowhouse and the church, halfway up the darkened lane, the hedge was broken by a brushwood gate. When I was some yards short of it I heard it creak, it opened a little and three figures stepped across my path. I slid to a halt, my heart beating wildly.

'Where you going, witch-lover?'

The reeve's son stood in the centre of the lane, a sickle in one hand. His companions grasped lengths of wood. I said nothing, but lowered the pitchfork slowly from my shoulder.

'I'd put that down if I was you.'

I stood still and remained silent. One of the other youths was edging around to the right, hoping to slip behind me. I swung the pitchfork and it brushed the hedge on both sides. He stopped.

'Put it down and you can pass. It's the witch we want, not you.'

I swung again and stepped forward and they moved back a pace. I had the advantage. The realisation released a surge of energy and I jabbed forward and took another step. If I could reach the gate I could escape into the field. The reeve's son was glancing to left and right as all three sought their opening but I kept swinging and jabbing and they were forced to retreat before me.

As I drew level with the gate I began to sidestep towards it, still swinging. The reeve's son now had his arm extended and was slashing the air with the sickle in an attempt to distract me. In a moment there would be enough space for the second youth to back along the far hedge and slip onto my other side. The third was now raising his arm to hurl his club . . .

I jumped sideways for the gap as the timber sailed past my head and crashed into the hedge but my jerkin snagged on the brush-wood and, as I fumbled to free myself, all three sprang forward. Glinting dully, the sickle arced across my vision. I ducked and it clattered harmlessly off the top of the gate. The second youth swung his club and it caught me round the ribs. I staggered backwards, thrusting down the long shaft of the pitchfork behind me to steady myself. I felt the handle lodge in the mud and braced myself on it, recovering my balance, then came a second impact, twisting it violently in my hands as the reeve's son, lunging for my waist and unable to stop himself, went head down onto its point.

He shrieked as the tines slid into his flesh, one beneath his cheekbone and the other through the base of his throat. I released the pitchfork and it sprang upright as he collapsed, gurgling, on the ground.

For a moment the other two youths and I stood motionless as the reeve's son wriggled and clawed at the neck of the pitchfork which quivered over him as if it were alive. Then, quite suddenly, his back arched, he shuddered and fell limp. His companions turned and fled down the lane as I sank to my knees and vomited into the blood-spattered snow.

I remained there, my mind in the grip of a cold, hard paralysis, until at length the gathering clamour from the village intruded, forcing me to rise. I glanced back to see the torchlight moving slowly but steadily towards the common. Giving the corpse as wide a berth as I could, I stepped back into the lane and stumbled on to the manor, where I fled up the steps to the steward's apartment and pounded at the locked door.

Footsteps approached, the bolt was drawn back and the door opened a little. The steward's wife looked out. She ushered me inside without a word and listened as I endeavoured to relate what had happened, but the horror of it seemed to have dislodged my thoughts and they tumbled out in breathless, nonsensical gabble.

She led me to the table and sat me down, brought me a cup of water and eventually I began to regain my coherence. Roland's aunt was sitting at the fireside. As I spoke she stared fixedly into the flames, clasping and unclasping her hands. When I had finished, the steward's wife looked at me for a long while. Her round, pleasant features were knitted with concern and the lively eyes had grown distant, reaching uncertainly beyond me as if for the answer to an impossible question. I realised that she was utterly at a loss.

'Where's Roland?' I had began to shake and thought I might vomit again. She nodded towards the sleeping-quarters.

'And the steward?'

'He went to see the reeve – to try and put a stop to all this.' She spoke quietly, deliberately, but she could not conceal the tremor in her voice.

Entangled in my own nightmare, I had not mentioned the approaching torches. She put her hand to her mouth in dismay as I told her about them, then said: 'you must take Roland and his aunt away from here . . . no, wait, Roland will not be able to travel – not yet . . . I know, take them into the manor and hide there . . . if the reeve asks, I'll say something . . . that I have sent them away . . . something . . . you must go, Creb, now. They'll find the body on the way here, if they do not already know . . . and then . . .' She looked away for a moment. 'They'll find some way to bring Roland into it . . . I know they will. Go into the manor. Another few hours' rest – that is all Roland needs . . . then, if you have to, go into the forest . . .'

She turned abruptly and went into the sleeping-quarters. A short while later she reappeared with Roland. He was extremely pale and looked drunk with sleep. I was not certain he had fully grasped what was happening and I went towards him but he waved me away and nodded solemnly. He took his mother's arm and walked with her to the door, then stopped and embraced her. I went to his aunt, who rose reluctantly from the table and allowed me to steer her, as if in a trance, across the room.

At the door, Roland's mother drew me to one side and said softly: 'If anything should happen, Creb, go with her to the town . . . if you can . . . if the pestilence has passed. I know she says little – but she is a good woman . . . and she has some money.' She took my hand in both of hers and squeezed it, then turned to face the fire.

Roland, his aunt and I left the storeroom and hurried across the courtyard. The torchlight was flickering up the lane, spilling over the hedgerows and casting bloody shadows across the snow-mottled fields. The voices had become less clamorous now and the predominant sound was the steady, purposeful tramp of feet.

We entered the hall, pulled the great door to behind us and stood for a moment in the darkness.

'Our lord's chamber,' whispered Roland, his voice trembling. 'It gives onto the courtyard . . . and there's a way down to a door at the back.'

We climbed a spiral stairway leading from the hall and entered the chamber. I went to the window, unlatched the heavy wooden shutters and eased them apart. The damp evening air stole in, bringing with it enough light to reveal the room's single furnishing: the frame of a large, canopied bed from which the mattress and hangings had been removed. For some minutes Roland's aunt hovered like a moth in the doorway. Then she allowed Roland to lead her into the room where she made for the bed and sat down upon it, her eyes closed and her hands clenched tightly in her lap.

Roland and I stood side by side, peering through the chink in the shutters. To the left rose the low, shadowy outline of the stables, kennels and mews. Directly opposite, forming the third side of the cobbled courtyard, stood the storehouse, the long ridge of its thatched roof thrown into flickering relief by the approaching torchlight.

Shortly, from around the end of the storehouse spread a dull glow, broken by grotesque, elongated shadows. I sensed Roland stiffen at my side. A moment later the party rounded the corner from the lane and clattered into the courtyard. It was led by the reeve and his henchmen. Twenty or thirty villagers trooped behind, some carrying torches, others an assortment of scythes, billhooks, staves and axes.

'Oh . . .' Roland gave a little choke of dismay. Beside the reeve, pinioned by two large, swaggering youths, trudged the steward. His hat was gone. White hair trailed untidily across his forehead and down over one swollen, half-closed eye. His gown was torn and mud-stained and although he was doing his best to carry himself erect, it was clear that he was dazed and considerably shaken by whatever had taken place to deprive him of his dignity.

As they made their way through the courtyard, the reeve glanced towards the entrance to the storeroom and raised his hand. Roland clutched my arm as the clattering of feet died away and out of the storeroom stepped his mother. She took a few paces across the cobbles and stopped in front of the reeve.

'What have you done to my husband?' Her voice was almost steady but I could see that one hand was plucking at her skirts.

The reeve's eyes strayed restlessly around the torchlit courtyard. 'Nothing to what I'll do with the witch.' There was such scarcely controlled violence in his voice that the steward's wife flinched.

'Witch?'

'Your son, Mistress Steward, he's a witch – possessed. He makes utterances.' The crowd muttered approvingly. Roland gasped and his fingers sank into my flesh as the reeve went on: 'It's well known he made the icicle fall and kill the cow . . . and helped the wheelwright escape . . . and now he's plotting with Satan against the innocents of this village, each and every one of us. He has the lad Creb in league with him . . .' his voice was rising 'and that

. . . witch-lover . . . tonight . . . has killed my son . . . stabbed him . . . with a pitchfork . . .' He paused as something almost feral invaded his features, then, with quiet menace, concluded: 'Devil's work indeed. And we want them – both of 'em.'

'But I have told you – repeatedly – uhh . . .' The steward doubled over, mouth hanging open, as one of his captors elbowed him in the midriff.

His wife winced and raised her voice angrily. 'Let him alone! He's done you no harm.'

She took a couple of paces towards him. One of the youths stepped forward and pushed her in the chest. She toppled backwards onto the cobbles with a grunt of pain. The steward struggled to break free and the reeve nodded to the second youth who suddenly released his grip and the steward went sprawling to the ground beside his wife.

There was laughter as the two of them struggled to their feet and the steward, turning to face the crowd, placed an arm protectively around his wife's shoulders.

'Now – where'll we find 'em?' The reeve's eyes were flickering around the courtyard again. 'Spare us the trouble of looking and maybe we won't bother you no more.'

There was a brief silence before the steward's wife answered hesitantly: 'I . . . I am sorry about your son, Master Reeve. We both are – truly sorry.' The steward nodded gravely. 'But I cannot speak for the young man, Creb. I've not seen him since late afternoon. I do not believe he is here and if I knew where he was, you can be sure I would tell you.' She held out her hands in a gesture of conciliation.

'He will have fled into the forest, I fear,' interjected the steward, picking up his cue.

The reeve squinted suspiciously. 'Hmm . . . then for now we'll settle for the witch.'

'No, by God!'

The steward made to step forward but his wife spoke again, her voice sounding small and frail: 'I'm . . . afraid not, Master Reeve. You see . . . I sent my son away . . . with his aunt. They left several hours ago.'

There was a groan of disappointment and the reeve's eyes narrowed still further. The steward, meanwhile, gave his wife a shocked glance and she nodded in confirmation.

'In any case,' she continued, a desperate tremor creeping into her voice, 'can you not see, all of you, that he is no more a witch than you or I? He is afflicted, it's true, but not possessed. He is an innocent lad who has fits . . . and sometimes he knows not what he is saying. That is all. And you should feel compassion for him, not hatred.' She had grown very pale and had begun to tremble. 'Can you not see that? What good would it do to harm him?'

From the darkness behind me came a mortified intake of breath, while down in the courtyard heads nodded here and there in sympathy. But the reeve turned to reassert himself.

'So the witch and his aunt have gone – and if they catch the pestilence that's their misfortune and good riddance. But murder's been done tonight – to my own flesh and blood – and justice is due. Not steward's justice, wheelwright's justice – but proper eye-for-an-eye, tooth-for-a-tooth justice. And I say there are places here where a murderer could hide before he'd go running off into the forest . . .'

He paused with arms outstretched, every inch the demagogue now, then nodded in satisfaction as someone cried: 'Try the church!'

Roland shrank from the window as the crowd began to murmur and press forward, spilling sideways and moving up the courtyard until the steward and his wife were surrounded.

'The hall,' came another voice from the crowd.

A predatory glint entered the reeve's eye. 'Friends,' he called out, 'friends, why not start nearer at hand?'

He turned, pointing theatrically towards the steward's chamber and moved forward. The crowd cheered and surged in behind him, engulfing the steward and his wife who were immediately lost to view.

I turned away and took Roland's arm.

'Where will we go?' he asked faintly.

'To the forest . . .'

There was a low moan from his aunt. She had begun to shake her head. Her face was bloodless, her fingers clenched to the frame of the bed.

'We *must* go,' I whispered.

Now she closed her eyes and her lips began to move in silent prayer.

I bent close to her. 'There's nothing to fear in the forest. I've been there. We'll be safe.'

She merely shook her head and moaned again, tightening her grip on the bed.

I stood up and looked at Roland: 'It's not her they want . . .'

Above the muted babble of voices, the sharp sound of breakage carried across the courtyard. It was followed by a woman's cry. I grasped Roland by the wrist and drew him towards the far end of the room. As I opened the door he paused, glancing over his shoulder, but he said nothing, pulled the door to and followed me headlong down the steps.

We groped our way down the darkened passage which ran along the back of the hall and reached the outer door. As I unlocked it there was a baying cheer from the courtyard. Roland hesitated for a second but I tugged at him and then we were racing out across the darkened deer park towards the looming bulk of the forest.

Despite the need to run as I had never run before, to release all the horror and fear of the previous hours, I knew, even before we reached the trees, that we would not be pursued.

The reeve had already obtained what he wanted – what everyone had known he had wanted from the very start: the uppermost hand in the village. In Roland, he had found the means to undermine the steward; in me, an unexpected furtherance of his cause; in the steward and his wife, a pair of defenceless scapegoats for the disorder he himself had engineered – hostages, perhaps, to my capture, which he would ensure did not take place. I pictured the eyes and mouth as I had last seen them, cruel and venal, and thought that he would have readily traded his son's life for control of the manor and its booty.

As we gained the cover of the forest Roland halted, gasping, and turned. Solitary torches pricked the darkness, moving fitfully around the manor but none towards us.

'I don't understand . . . they're not coming . . .'

'They'll be searching the outbuildings first . . . we must go on.'

Safe though we almost certainly were, I had the feeling that there would soon be much unpleasantness back there in the courtyard, if it had not already begun. For that reason alone I wished Roland – both of us – far away on this night.

We continued in silence. The sky was overcast, the darkness

intense and it was all we could do to thread a path between the trees and bushes. I had the charcoal-burner's hut vaguely in mind for shelter but could only guess at our direction. Shortly we slowed to a trot and then, as Roland's breathing became more and more laboured, to walking pace. A little while later Roland stopped and sank to the ground with an apologetic shake of his head. There was a large fallen tree nearby. I crawled through the tangle of its branches and found, beneath the trunk itself, a patch of dryish ground. I called to Roland to follow me but there was no reply. Scrambling out again, I found him sound asleep where he sat. I woke him gently and led him on hands and knees into the space beneath the tree where, at my bidding, he stretched out and allowed me to cover him with armfuls of leaves and other dead stuff to keep off the worst of the chill.

He wriggled to make himself comfortable, then sighed deeply and said: 'Creb?'

'Yes.'

'What did he intend by "utterances"?'

I hesitated.

'Is is that I . . . that I say things?'

Some instinct told me to be truthful now, however inappropriate the moment might seem. 'Yes. You do. When you've a fit coming.'

'Have you heard me say these things?'

'I have.'

'What are they? What do I say?'

I paused. 'No more than you know already.'

'Oh . . .'

A few moments later he was asleep again. I thanked God for the fit which, together with the drama at the manor and the exertion of our flight, had left him quite exhausted. He would be better able to deal with our predicament in the morning.

But what precisely was the nature of our predicament? I lay back and strove to blot out the recurring images of the quivering pitchfork as I imagined the way things might be resolving themselves in the village. Not to my advantage – of that I could be sure. But might the steward recover his authority? It would be well-nigh miraculous, but if he did, Roland at least would be able to return. Despite my deep misgivings I felt a growing obligation to make certain. At length I crawled out from our shelter and, taking good note of my route, made my way back to the edge of the deer park and paused in the shadow of the trees.

The manor was dark and silent. A dog barked distantly. Straining my ears, I caught the faint sound of voices but they were far off in the village. Cautiously I crossed the deer park and crept around the end of the kitchen to enter the courtyard. It was deserted but all around, doors hung open and a glance through the first, into the kitchen larder, revealed that the place had been ransacked.

I tiptoed across the cobbles and into the storeroom. It too had been overturned, the floor strewn with spilt grain. I climbed the steps and peered into the steward's quarters, wincing as I saw what had been done to them. But there was no sign of the steward or his wife.

I recrossed the courtyard and entered the hall to be struck at once by the fragrance of herbs – much sharper, it seemed, than when Roland and I had laid them, the previous afternoon. As my eyes grew accustomed to the darkness, the reason became apparent. Undulating with untidy mounds of rushes, crushed and trampled, the floor resembled that of a threshing-room in which three darker and more substantial shapes at first suggested the exhausted forms of labourers who had dropped at their work. The steward's sister was on her back, not far from the foot of the stairs. Her gown had been wrenched up above her waist and her naked belly and the upper parts of her thighs were livid with bruises. The steward and his wife lay in one corner of the great chamber, as if they had been brought to bay there, she with her face to the ground and he sprawled protectively across her, his crown misshapen and piebald with blood.

For some time I stood, quite numbed, and stared into the gloom. Then, feeling that I must do something for them, I cast around for a shroud. By the door someone had dropped a shawl. I retrieved it and laid it across the heads of the steward and his wife. There was nothing else with which to cover Roland's aunt but I pulled down her clothes and straightened them as best I could. There was a soft clinking as I did so and after a moment's search, my fingers closed on a handful of small flat discs sewn into the hem of her gown. Dismissing a flutter of conscience, I tore open the fabric and withdraw a dozen coins which I knotted into a tight bundle in my shirt-tail. Then I quit the hall and foraged hurriedly through the adjacent buildings, emerging with a small sack into which I had stuffed a bag of flour, some shrivelled vegetables, a bottle of sour wine and an ancient cooking pot.

As I left the courtyard I paused and glanced down the lane toward the village, momentarily overwhelmed by the desire to slip back to our cottage and persuade my mother and father to come with me, or at least say a decent farewell to them. But reason prevailed and I turned reluctantly for the forest.

When I reached the thicket, Roland was awake. He raised his head on one arm as I lay down beside him and said: 'You've been back to the manor.'

I patted the sack. 'Food.'

'For how long?'

'A day or two – what I could find.'

He paused. 'It won't be enough. We're not going back . . .'

'Well . . .' I began, but he caught my wrist and said quietly: 'Creb, I know. They're dead.'

I sensed him stiff and silent in the darkness beside me and prepared myself to comfort him. But after some moments he laid his head on the ground again and, in a voice thick with fatigue, said: 'Swear you'll never tell me the manner of it.'

'I swear it,' I said.

PART TWO

Seven

tiff, cold and hungry, we crawled out from under the tree at first light and stood chafing our limbs beneath a cloudless, eggshell sky. A sharpness had returned to the air and I knew that we would need decent shelter for the night to come. We also needed to rest before setting off for wherever it was we would ultimately go. Roland was pale, silent and sluggish and although I could not tell what he was thinking, I guessed that his expression – slackened by shock and grief – must mirror my own.

We set off through a forest still wreathed with wisps of mist. Gradually they dispersed in the rising sun but Roland and I remained beyond its touch. We spoke hardly a word to one another as we made our way in search of the charcoal-burner's hut. I could not rid myself of the thought that only a few hours ago I had killed a person and although my mind told me it had been in self-defence, and an accident at that, some other part of my being insisted I had done a dreadful wrong, an abhorrence against nature. I had extinguished a life.

We reached the clearing after an hour or so. As we approached the hut, Roland grimaced and put his hand over his mouth. I told him to wait outside and ducked through the entrance, gagging as the putrid stench caught me in the back of the throat. One glance at the monk's corpse was enough to confirm my suspicions. I stumbled out again, backwards, and ran to the other side of the clearing to gulp in air.

Perhaps I had become more pragmatic over the last couple of days; perhaps I was simply brutalised, my finer sensibilities driven underground. In any event, I quickly dismissed any notion of trying to bury the monk properly. It was merely a question of how to dispose of the corpse – and particularly the smell – most quickly and efficiently.

In the end, since we had nothing but our bare hands with which to shovel, we decided it would be easiest to burn him. Roland looked green at the prospect as together we drew deep breaths, entered the hut, grasped the monk each by an ankle and heaved. He was as light as his emaciated appearance had suggested and without too much trouble we dragged him out, across the clearing and into one of the charcoal-pits. Roland walked away unsteadily and puked into a bush as I moved back some distance from the pit and, with the monk's own flint, started a fire.

Once it had caught well, we armed ourselves with branches and were able to fork enough burning material into the pit to build up a decent blaze. Then, as a plume of greasy, foul-smelling smoke drifted up over the clearing, I sent Roland off with the pot to find water and rearranged the embers of the smaller fire for cooking.

On his return we kneaded flour and water into little cakes and laid them amongst the ashes. When we retrieved them some time later they were lumpy, tasteless and only half cooked. A leathery carrot or two and a swig of sour wine gingered our palates a little, but it was not a meal to savour.

After we had eaten we lay by the fire and talked for a while of inconsequential things, neither of us able yet to digest fully what had gone before or to consider what lay ahead. At length we both fell asleep.

For the rest of the day we dozed intermittently, leaving the fireside only to forage for more fuel. Then with the lengthening shadows came the first bite of frost and we moved into the hut, where a fresh fire quickly rid the place of the last vestiges of the monk's presence. I prised open a hole in the roof to let out some of the smoke and once our eyes had stopped watering we made another meal of dough cakes and a boiled cabbage stalk.

'What will we do?' asked Roland eventually, pursing his lips and laying down the wine bottle. Although he seemed physically restored by the sleep, his voice was still flat and his eyes dull.

'See what goes on beyond the forest,' I replied, thinking as I spoke. 'If the pestilence has passed, we'll get work. If not, we'll come back here and make do . . . somehow . . .'

'Do you think it has passed?'

I shrugged. 'Don't even know what brought it. I thought maybe it was in the air, like rain or dust . . . but then it would've got here long ago, unless someone was keeping it away on purpose. Can't see why they'd bother with that, though . . .'

Roland went quiet and still. 'I'll likely be damned for saying this, Creb, but . . . I'm beginning to doubt whether Someone exists. How can I believe it any longer? How can I?' He shuddered and put his head in his hands. 'I'm accustomed to knowing things, Creb . . . understanding things . . . sometimes before I've even given them thought . . . but now it's all contrary . . . I understand nothing, no matter how hard I think . . .' He paused, his eyes moistening. 'My parents . . . they were decent people . . . did they really deserve to . . .?'

Much as I longed to comfort him, I had no answer to the question. We sat in wretched silence until at last, mercifully, his own words returned to me and I was able to reply: 'Perhaps deserve isn't a part of it. Perhaps it was just their time.'

He stared into the fire for a long while and finally gave the slightest nod. 'What else is there to think?' He sniffed and poked at the embers with a stick. 'Do you believe, Creb?'

'I don't know . . . there has to be something . . . couldn't make sense of what's been going on otherwise . . . but what is it . . . now there's a question.' I paused. 'One thing's for certain though. Whatever it is, it doesn't always do what you expect . . .'

On an inspiration I began to tell him the story of my previous adventure in the forest. By the time I had finished he had brightened a little. Eager to sustain the new mood, I removed the parchment from my shirt.

'This is what I was going to show you.' I passed it to him. 'It was in the monk's satchel.'

At another time, in another place, Roland might have made a fine scholar, perhaps even a great one. He would have had to have learned to approach his conclusions more judiciously, to curb the wilder excesses of his imagination, but the essentials were present – an infinite curiosity, considerable mental dexterity and an unusually sharp intuition. As it was, however, these assets now conspired to set us on a path which I could never, in a millenium, have begun to guess at.

He spread the parchment on the ground at his knees and scrutinised it briefly before declaring that it was a map, a form of description of the countryside which told a traveller how to get from one place to another. Pilgrims used them to show the way to holy places, he said. But this one was odd because the names of the places it illustrated were absent and if the features were unidentifiable, how then was it to be used? He gave a sigh of frustration and began to name aloud the seven places which were marked distinctly and at varying intervals along the road: a castle, a gibbet, a round hump which he took to be a hill, a church, an anvil, a well and what appeared to be a cave in a mountainside. Assuming these places existed, he said, and in that order, it was almost certainly the representation of a journey. But did this journey end at the cave or the castle? And even knowing that, there was still no indication as to where either might be, nor, more importantly, any clue to the significance of this particular journey.

He looked up at me. 'Was this all he had?'

'No, there were pages of writing . . . wait . . .' I turned to grope in the darkness behind me. 'Here,' I produced the satchel and passed him the sheaf of parchments. 'Can you read it?'

He held up his hand and began to scan the pages. I placed more wood on the fire and the flames leapt up, lighting the intense concentration on his face. After some while he laid them down and shook his head. It was mostly in Latin, he said, and the rest seemed unintelligible. He chose a passage and read laboriously: 'What is below . . . is like that which is above . . . and what is above . . . is like that which is below . . . to acc . . . accomplish . . . the miracles of one thing.' He rolled his eyes in incomprehension and I began to see my parchment dream crumbling to ash.

Trying hard to conceal my disappointment, I asked him if he thought there might be anything in the drawings surrounding the text. He ran his hand through his hair in a gesture that reminded me momentarily of his father and began to turn the pages again with a frown of deepening perplexity, as his finger came to rest in turn on a sun with a crown, a fish, a dragon . . . Reaching the last page, however, he paused thoughtfully then returned to the first, studied it intently and nodded, moved on to the second, the third and so again to the end.

'Give me the map.'

I passed it to him.

He was silent for a moment, his gaze flickering between the drawing and the sheets of parchment. When he looked up again the vitality had returned to his eyes. It was possible, he said, that there was some correspondence between the seven principal features on the map and seven particular words which occurred in the text – particular not only because they had been illuminated but also because the illumination appeared to have taken place subsequent to the original drafting of the text. I leant across and in the flickering light could see clearly enough that here and there was a single word which had been minutely and intricately embellished – as an afterthought, it appeared, since the detail of the embellishment crowded back against the end of the preceding word and spread up and down to touch both the lines above and below. He read them out as he went: chivalry, treachery, earth, air, fire, water, transmutation.

'Trans-mu-tation,' he repeated, sounding increasingly animated. 'Do you remember? When lead is turned to gold – the process?'

A vague memory stirred as he went on to explain that the words did not appear at the points in the text which would normally have been illuminated. Instead, they sprang off the page from the middle of sentences, the ends of lines and other places which, he assured me, were most unorthodox. As I grappled with the implications of this, Roland began attempting to match them to the figures in the map. It was no more than a hunch, he said, but worth trying for all that. He started with the well. Water, presumably. The next one, the cave. He scratched his head. Earth . . .? Maybe. Then back a couple to the anvil. His fingers began drumming a tattoo. Chivalry, treachery, fire . . . *fire!* Of course, a smithy. The gibbet next . . . that had to be treachery. He paused to recapitulate: something, treachery, something, something, fire, water and earth; then paused again and flicked back through the pages of text.

'Mother of God, Creb, they're even all in the proper order! Listen!' He stabbed the map repeatedly with his forefinger. 'Castle – chivalry. Gibbet – treachery. Hill – I was wrong about that – it's earth. The church steeple – air. Anvil – fire. Well – water. That leaves the cave for transmutation.' He looked up at me with an intensity that made me shiver. 'Do you know what I believe this map does, Creb? I believe it tells you where to find the Philosopher's Stone!'

I gazed at him blankly.

'The Philosopher's Stone. The agent of transmutation. Lead to gold. The very stuff of alchemy!'

He turned breathlessly to the other figures on the written pages and began to point out what he now claimed to recognise as alchemists' instruments: a bellows for the furnace, tongs, a crucible in which to melt the lead, and so on. His eyes darted to the text again and he read: 'Visit the inward parts of the earth and there, with con-tem-plation, find the hidden stone.'

He got to his knees and grasped my shoulders, his face alight. 'Do you understand, Creb? Follow the map from the castle to the cave – there to find the Philosopher's Stone. This is remarkable! Extraordinary! It could mean wealth, Creb, fantastic wealth, beyond the dreams of avarice . . .'

By now I was quite lost. I shook my head.

Roland rocked back on his heels and patiently began to explain that each word had a second, hidden meaning. Chivalry, he said, stood for perfection, treachery for baseness. Earth, air, fire and water were the four elements from which everything in the universe was composed. Transmutation, finally, was what occurred when a base substance, like lead, was turned into something perfect, like gold. It so happened that these words were also represented on the journey, in the form of real features along the way – like the Stations of the Cross, for instance.

The logic of it was evidently as clear as daylight to him, but to my more pedestrian mind the whole thing was shrouded in a dense fog of unpronounceable words and obscure symbols. I enquired hesitantly why our monk, if he had known all this, had ended up here, dead? Why not rich?

Roland shrugged and replied that it was obvious, was it not? The monk had not yet been to the cave. He had been setting out on his journey, not returning from it. Then he gave a sudden frown and asked me whether there had been anything else in the satchel. I nodded slowly, remembering all at once what had tumbled from it. I put my hand inside, felt for the strange rough object and pulled it out with a flourish.

Roland squinted at it, then burst out laughing. 'Pumice! It comes from volcanoes. Scribes use it to clean parchment – make it nice and smooth so the ink doesn't blotch. No, Creb. He would never have transmuted anything with that!'

I asked him sharply to tell me then, since he appeared to know everything else, where the monk had been heading for and why he had only come so far. But Roland's enthusiasm seemed to leave him impervious to sarcasm and he admitted quite cheerfully that he had no answer to the first part of my question – yet. As for the second part, it was again perfectly obvious. The pestilence had driven the monk into the forest and he was incapable of looking after himself. They led a soft life in the monasteries, he added with an air of authority, then returned his attention to the manuscript, squinting at it in the firelight. I watched him sprawling on the ground, shoulders askew and head down, rapt in boyish concentration. At that moment I did not know whether I was infuriated by his intelligence and confidence, or infatuated with it; aghast at his ability to shrug off his grief, his despair, so completely, or delighted by it.

'Aha!' A sheet of parchment was thrust under my nose. 'And there, with con-tem-plation, we find the hidden castle!'

He laid it down beside me, nodding appreciatively, and told me to look at the illumination of *chivalry*.

I did and, needless to say, saw nothing.

'The pennant, Creb. Look closely. The pennant!'

Fluttering from the tip of the letter L was a pennant, while still closer scrutiny revealed that the letters A and R, on either side, appeared to be superimposed on the battlements of a castle. The pennant was emblazoned with a minute boar's head. It was wearing what looked like a wreath around its neck and a ball on the end of each tusk. I was still peering at it, wondering what its significance could be, when another sheet was thrust triumphantly on top of it and I was invited to inspect the embellishment around the word *transmutation*. Lo and behold, teetering on the summit of the letter A, there was the tiny figure of a goat, horned and bearded. Sounding distinctly pleased with himself, Roland explained that of all the seven words, only these two, the first and last, bore anything in their illumination which was not entirely abstract. All the rest were surrounded simply by swirls and curlicues. But *chivalry* had part of a castle and a little flag whose emblem – the boar – belonged to a lord who had visited the manor to hunt. He remembered it because he and our lord's son had named the visitor

Porker and been thrashed for it. He did not know where the castle was, he said, but it was unlikely to be far away since no one in their senses would have travelled any distance to visit our manor. As for the goat, he concluded, it had to be a clue to the journey's destination which would only become apparent as we came nearer to it.

I found myself beginning to yield, perhaps a little reluctantly, to the force of his conviction but there were still things that puzzled me.

'When we find this stone,' I asked him, 'what will we do? Can you – make it work?'

Roland waved dismissively. 'I can learn. Or we might sell it.' He reached for the bottle, took a swig and spluttered. 'If I'm to be an orphan, Creb, I should like to be a wealthy one.'

What was I to make of it? If he was right, this exceeded my wildest anticipation of the kind of knowledge that the little drawing might bring, but at the same time, in some indeterminate way, I felt disconcerted. It seemed as if something almost mystical had been laid over the landscape, veiling the solid realities of timber and stone, grass and earth. What my instinct told me we needed at this moment was some measure of certainty, something we could grasp. But, on the other hand, we had nothing better, and it would lend us a purpose . . .

'Well?' He looked across at me, grey eyes gleaming in the firelight.

I hesitated for a moment, before asking finally: 'If he wanted his journey secret, why put clues in the map? Why make a map at all, come to that?'

Roland stretched and affected a long, weary sigh at which, this time, I could only smile, for I was certain that at last he was flummoxed. A moment later, however, he was rattling off a fluent explanation, concocted as he went along, no doubt, but camouflaged with all his usual zest. The gist of it was that the monk had not drawn the map himself but had come by it in some other way, and only after studying it had he realised that its features were open to symbolic interpretations echoed in his own writings. Then, in a transport of joy at his discovery, or perhaps, more prosaically, because he had a bad memory, he had sharpened his quill, mixed up the illuminating inks and, well – here we were . . .

He cocked his head cheerfully to one side and looked at me. 'So – what do you say?'

By now there was only one answer.

The frost that night was severe. Curled beside the fire like hedgehogs, we slept fitfully, waking every hour or so to put on more wood. Early next morning we left the hut and set off for the easterly edge of the forest. In the rising sun the air was crisp and still and brilliant with the sparkle of rime.

Restored by the inactivity of the previous day, my mind had sprung alert the minute I awoke; it reeled and groaned with the chaos we had left behind us and the uncertainty of what lay ahead. But after a while the walking seemed to quieten it and I found myself slipping into an unfamiliar state in which I caught glimpses of some place to which I had never been before. There, it was warm and sunny and peaceful and I was content. All my anxiety and confusion had gone. I knew that it was distant, perhaps unattainably so, perhaps even a mere mirage, but I also knew that so long as I could hold it in my mind's eye, the uncertainties along the way became tolerable. It lent an extra sense of purpose to our journey through the forest and even a glimmer of meaning to the ignorance and brutality that had driven us here. It was no more than simple optimism, a belief in the future, but an emotion which had so far figured insignificantly in my life, nonetheless. And what seemed especially curious was that it should surface now, when things had never been more precarious.

I glanced at Roland, walking beside me with the monk's satchel slung from his shoulder. The cold had brought colour to his cheeks and now he was alert once more, glancing keenly at the frosted features of the forest around us. What sustained him, I wondered. Was it the thought of the unpronounceable stone? Or was' it, despite his outburst of the previous evening, that uncanny ability to know and accept?

'The icicle's been on my mind,' he said some time later, as the glitter of a frozen waterfall drew us to pause and drink. 'I should like to know why it fell. Something did it – and it wasn't witchery.' He gave a self-deprecating grin.

I gasped at the iciness of the water trickling down my chin. 'My father didn't hold for mysteries. "There'll be an answer – always is." That's what he said.'

'He was right.' Roland nodded pensively. 'Perhaps it became too heavy . . . or a bird flew into it . . .'

'Would an alchemist know?'

'Mmm . . .' He flung out his arms. 'Think of the knowledge they must have, Creb! How the stars stay up . . . why oliphaunts have long noses . . . what love really is . . . how you come by it . . . what brings pestilence . . .'

'If there is pestilence . . .'

He chattered on as we walked, describing his own extraordinary vision of knowledge. Captivated by the colour and cadence of his words, I sauntered beside him in a pleasant dream as he painted glowing images in my mind. Every new thing learnt, he said, added a little to a person's stature. But it was not a matter of growth as we usually understood it, since it occurred downwards and upwards at the same time: down into the earth and up into the sky. And since it all took place within the spirit, not the physical frame, the growth was not apparent from the outside. Yet the more a person learned, the deeper he became rooted in the substance of the earth and the higher he reached towards the heavens, until the moment arrived when there was no more to absorb and his spirit spanned the gulf between fundament and firmament. Then he not only knew everything but also became everything: rock and soil, wind and rain, forest and river, steel and leather, thunder and lightning, chaff and grain, wheel and anvil, moon and stars . . . If this was the reward of alchemy, I thought, then I would happily forego the gold.

An hour or so later the sun was abruptly extinguished as a long, sharp ridge rose across our path, blotting out half the sky and forcing us to follow the stream that wandered at its foot through a woodland suddenly dense and shadowy. For some while we walked close together and in silence as the ridge continued to curve above us, deflecting us almost intentionally, it seemed, from the direction in which we wished to go. Then, equally abruptly, it came to an end in a ragged, lichen-speckled cliff. The forest thinned and sunlight flooded in once more. Three hundred paces further on, it ended altogether.

Roland and I stood in the shelter of the last trees and stared at the open countryside beyond. At our feet, the ground fell away in the sweep of a broad, steep meadow. Overhead, a cloudless winter sky arched infinitely across a crumpled landscape of grassy folds and wooded slopes. Here, the horizon sprang sharply into focus with the summit of a nearby knoll and, there, swam back to the gentle serrations of inconceivably distant, smoke-blue hills. This was the unknown, a vastness as intimidating as it was liberating.

After a long time I turned to check the position of the sun and when I turned back again Roland was intently scanning the middle distance.

'Curious . . .' he said, frowning.

'What?'

'No smoke.'

'So?'

'Well. It's clear. We can see for miles. There must be villages out yonder.'

I nodded, picturing the slim grey threads that hung in the air above the cottages on a still winter's day.

'So why no smoke?'

Now the puzzlement was laced with an undertone of apprehension. I began searching for some plausible solution, then thought better of it and asked him directly if it was the pestilence that troubled him. He replied with an uncharacteristically diffident shrug.

'We must know, Roland. Sooner, the better.'

He nodded solemnly.

'Maybe there's none.'

He nodded again, then made a little grimace. 'It frightens me – sickness. You can't see it. Can't touch it. Don't know where it's coming from – or when.' He looked down. 'I'm sorry, Creb. I'm not the bravest of comrades.'

'You'll do for me,' I said, smiling. 'Come on.

By my rudimentary reckoning we had emerged from the forest some miles south of the village at the end of the glen. Beyond the village there was purported to be a town but it was known only to those few of our own folk who had made the three-day round trip to market and back. For the rest of us, it was little more than a vague wrinkle in the imagination. But I had an idea that that was where we would be most likely to find directions for the castle. Roland agreed that since we did not know where this town was, however, and since we also needed food and proper shelter for the night to come, we should make first for the village.

I set out briskly along the crest of the meadow. A knot was gathering in my stomach and I was anxious to walk it off. I had gone a couple of hundred paces before I realised that Roland was not with me and I turned to see him trailing a good distance behind. Thinking it was reluctance that was making him drag his feet, I called out some encouragement and waited for him to catch up. But as he drew level I saw immediately that that was not the case, or perhaps only part of it. A concentrated flush had appeared on his pale cheeks, he was breathing hard and had begun to stumble slightly.

'I'm sorry,' he panted. 'You were going too fast.'

For a couple hours we continued along the edge of the forest at a more leisurely pace, climbing and dipping through wild meadowland with not a mark of cultivation or habitation to be seen. Then, around noon, a number of sheep appeared, ambling in single file across the face of a low hill ahead of us. A few minutes later we reached the brow of the hill and halted.

Beneath us lay a broad valley. In its centre, a good distance out from the edge of the forest, stood a small church, a barn, a cluster of animal houses and fifteen or twenty cottages. Bathed in winter sunlight, it seemed at first glance like a peaceful, sleepy place where nothing of any event had taken place for generations. Then came the realisation that it was devoid of movement.

For some time we waited, willing the emergence of people from the dwellings, the sound of voices, the barking of dogs. But there was nothing – just an eerie sense of desertion. When eventually we glanced at one another, Roland was unable to keep the anxiety from his face. I told him to wait for me and he nodded and sat down on the grass as I set off down the hill.

The floor of the valley was a patchwork of tillage and as I picked my way towards the cottages, a flock of several hundred pigeons rose noisily from a strip of kale. I glanced across to see row after row of ribbed green stalks rising naked from the earth with scarcely a leaf between them. In the corner of another strip, a pig rooted indolently at a pile of mouldy turnips and a little further on, as I stepped onto a track leading into the village, an ulcerous, brindled dog slunk away from the wheel of an abandoned cart.

I looked over towards the church, standing solidly apart from the cottages with its squat tower looming above a high yew hedge, and noticed for the first time a vague wisp of smoke, rising into the still midday air. Thankful that I was not obliged to enter the village, I headed towards it.

A voice began to make itself faintly heard and as I walked around the hedge, seeking the way in, it gained substance. It was high and thin, a child's piping, monotonous and repetitive:

> Bury them one by one,
> Bury them two by two,
> Bury them three by three,
> Under the chestnut tree.

I found the gate and entered the churchyard. On my left a ramshackle shelter of branches and pieces of sacking sprouted from the hedge. Before me stood a large headstone, raggedly fringed with uncut grass. At its foot smouldered a fire with the carcase of some small animal roasting over it on a makeshift spit. Beside it sat the singer, a filthy little girl of six or seven years. She had arranged the ash into rows of neat grey oblongs which she tapped with a stick, one after another, as she sang.

She glanced up at my approach, ran a grubby fist beneath her eyes, blinked and said: 'Are you the priest?'

I shook my head.

'Pa says a priest will come one day and say the right words.'

'I'm sure he will,' I said. 'Where is your pa?'

She pointed to the end of the church, then returned her attention to the ash and began to sing again. I walked to the corner, the macabre little rhyme piping insistently through the stillness, and looked around.

Partly shadowed by the tower, a field of fresh graves ran the length and a little more than the breadth of the church itself. All that remained of the turf was an irregular grid of rough green borders enclosing mound after mound of turned earth. A cloying, sickly smell hung over the place. At the far end, beneath the outer branches of a large chestnut, was a young man, naked above the waist save for a kerchief tied around his mouth and nose. He was shovelling soil into an open trench. An empty hand-cart stood beside him.

He looked across as I approached him but did not acknowledge me and kept on shovelling. I stopped a little short and for some moments stood watching him, uncertain what to say. Eventually I asked him if the pestilence had passed.

He turned to me, resting on his shovel, and wiped sweat from his forehead. The drooping shoulders, the trembling hands and bruised eyes told me that he was close to the limits of exhaustion. But there was something else in his look, an echo of some unimaginably dark place from which he had narrowly been delivered. It infected his expression with a hovering uncertainty that seemed not far short of madness.

'Passed?' He shrugged. 'It took 'em all.' There was a curious jerk to his speech, as if all the natural inflections had become misplaced.

'You're the only ones left?'

'Of them that didn't go elsewhere.'

'Why d'you stay?'

He frowned and gesticulated vaguely at the graves. 'To bury 'em, of course.'

'Are there many more?' I had been trying to count out of the corner of my eye but had given up after twenty.

'Oh yes.' He straightened and spat on his hands. 'Plenty more.' He turned to the grave and began to shovel again.

'Do you know how far it is to the town?' I asked.

He grunted something inaudible and tossed his head and I felt disinclined to press the question.

I made my way back to the gate to find the child emerging from the shelter.

'Why don't you sleep in the church?' I asked her.

For a moment she looked at me as if I was an imbecile, then replied: 'Because we're not fit.'

I left the church and headed back through the fields.

Halfway up the hill I turned round and looked out over the village. A solitary figure was making its way from the church to the cottages, wheeling a hand-cart before it.

Eight

t took us two days to reach the town but no more than a few hours to appreciate that in stepping forward into the unknown, we were merely trading one kind of chaos for another.

We should have anticipated it after my encounter in the churchyard, but we were still naïve enough to believe that with the passing of the pestilence, normality would have been instantly restored and we were quite unprepared for the disintegration that it had left in its wake. It would have been easier to comprehend, perhaps, had the cause been war; the countryside disfigured by the tramp and skirmish of conflicting armies. But the pestilence had perpetrated a subtle deception, striking at the marrow with no visible disturbance to the home.

The first village we came to was half empty, its surviving inhabitants gathered at the alehouse in an orgy of thanksgiving or grief – it was hard to tell which. Strangers, it became immediately clear, were not welcome. They were still less welcome at the next. We were driven away before we had even reached the first cottage. A little further on, the tracks and paths began to criss-cross one another more frequently; emaciated cattle and sheep browsed the ditches and hedgerows and we became gradually aware that a scattered army of the dispossessed was roaming the countryside – singly, in pairs, in small groups, some of them apparently quite aimless, others scavenging intently for whatever food the woods and meadows and untended fields might yield.

That night we passed in the open, huddled together with our backs to a stone wall, a fire in front of us and makeshift cudgels in our hands.

On the morning of the second day, dropping down into a narrow wooded valley, we stumbled on a hamlet hidden in a clearing. Only after producing the coins from my shirt-tail were we reluctantly admitted. The pestilence had passed them by, we were told. It was a miracle. But there was none of the euphoria associated with such events and the listlessness of the inhabitants, their dull strained faces, suggested that this miracle had cost them a good deal.

They were prepared to sell us a little food, however. I also enquired if they might have a donkey, for it had become increasingly apparent that Roland's condition did not lend itself to sustained exertion, and he had agreed reluctantly to the proposition that we might reach our destination more quickly if he did not have to walk all the way.

Two donkeys were produced. After a brief inspection he selected the older and more disagreeable looking of the two and promptly christened it Nebuchadnezzar. I raised an eyebrow, not even attempting to utter the word. It sounded quite implausibly obscure.

'An old king from the scriptures,' he explained, climbing onto his mount. 'He tried to roast three children of Judah in a fiery furnace but the Lord saved them.'

'And what has that to do with a donkey?'

He shrugged. 'It just seems right.'

Nothing seemed right about this brute, I thought, taking a closer look and marvelling at the perversity of Roland's choice. Mangy and mud-caked, with a filthy knot of bristle tipping its tail like the end of a rotting bell-pull, the creature stared implacably from beneath enormous eyelashes as I fumbled with the rope halter. Then, at the first hint of a tug, it effortlessly transferred all its weight to its hindquarters, dug its heels into the ground, bared large yellow teeth and brayed.

'No, no,' said Roland, trying at the same time to place his hands over his ears and grasp a handful of mane to prevent himself from slipping off backwards. 'You must talk to him. Encourage him.'

There would have been more pleasant conversation to be had from a dead dog, I thought, as I put my mouth close to the scrofulous head and whispered cajolingly: 'Come on, shite-ears. Time to work. Understand? Off we go.'

I tugged at the halter again and the brute responded with a

sudden lunge at my forearm. I withdrew it just in time to avoid the clack of gin-trap jaws.

Roland sat and cackled with laughter as I went off in search of a stick. I had to go some distance to find just what I wanted – supple enough to raise dust from the moth-eaten grey hide, thick enough to whittle into a good sharp point – and by the time I returned, they were trotting cheerfully towards the trees.

I caught up with them, out of breath and out of humour, and remonstrated with Roland for not waiting. He grinned, flexed his knees against the creature's sides and it halted obediently.

'You won't need that,' he said, pointing at my stick and starting to laugh again. 'Not unless you intend to ride him.'

'Ho, ho!' I tossed the stick away and asked him sourly where he had come to learn about donkeys.

'I didn't. But it's elementary. Treat them with respect and they'll do what you want. And you were disrespectful, Creb. You called him "shite-ears". He didn't take kindly to it.' The creature turned to me with a baleful look, peeled back its lips and expelled a string of greenish-grey froth from both nostrils.

Roland coughed and dabbed at his eyes, then scratched his mount affectionately on the head and set off again.

By noon that day the countryside had begun to level. In the late afternoon the track led us across a broad expanse of pasture towards a line of pollarded willows, their clusters of naked branches sticking out from the boles like filament against a watery sky. We approached the riverbank on which they stood to find a crowd of travellers awaiting their turn on the ferry – a wooden raft being poled from bank to bank by a muscular youth with features severely pitted from the pox and an air of deep resignation at the monotony of his task.

We tethered Nebuchadnezzar to a bush, then took our place at the back of the crowd. For some time we sat and listened to the conversations around us as a three-quarter moon climbed slowly into a pale sky and the duck whirred overhead on their evening flight upstream. The mood of this gathering was quite different to that in the villages, where the ranks had closed to protect the collective dignity. Here on the riverbank, a stranger could air his deep private wounds, knowing that in a little while he would continue anonymously on his way with his self-respect intact and something gained from his unburdening. There was even a sense of camaraderie in the telling of these tales of horror and deprivation, acts of courage and strange twists of fate and, although Roland and I remained on the edge of it, too shy to join in, we knew that we shared the common experience, that we had all been touched by something great and terrible. In a curious way it cheered us.

Our turn for the ferry came a little after sunset. It was to be the last crossing of the day, the ferryman declared. By now only half a dozen of us remained. The raft was large enough to take us all with the donkey and room to spare but nothing would persuade Nebuchadnezzar to set foot on the gently rocking wooden platform. The ferryman, who had already extracted an exorbitant fee for our passage and was clearly accustomed to this sort of thing, leant on his pole and yawned. Roland tugged at the halter, cajoling and cursing, whilst I, endeavouring to conceal my amusement, heaved at the grey rump which sank intractably ever closer to the ground.

Eventually the ferryman fished at his feet and tossed us a length of rope.

'Tie one end to 'alter and t'other to that there.' He indicated a ring attached to the edge of the raft, then stood back again with folded arms.

A stricken look entered Roland's face and he protested that Nebuchadnezzar would surely drown.

"Course he won't,' said the ferryman. 'You want to cross or not?'

I suggested that if we were to avoid spending another night in the open, we would be well advised to do as the ferryman said. Roland stared at the water for a moment, then gave me a peevish look and began to attach the rope to the halter, muttering to himself as he did so. Then, holding the other end, he jumped aboard and squatted down to tie it to the ring. As I joined him, the ferryman turned to face the far bank, hefting the pole in his hands. Satisfied that the knot was firm, Roland stood up and tapped the ferryman on the shoulder, then turned to stare gloomily at the bank. The donkey had now lowered his rump to the ground and was sitting with forelegs splayed, ears pricked and a distrustful look in his eye.

The ferryman gave a thrust with his pole, the raft slid away from the bank and the donkey lifted his head and brayed mournfully as the rope uncoiled, sagged for a moment over the water, then snapped taut, jerking him to his feet and cutting him off in full voice. But somewhere in the depths of Nebuchadnezzar's mangy grey soul lay a measureless reserve of obduracy. He slithered forward a few inches, dug in his heels, sat down again and the raft halted. The ferryman muttered an obscenity, the passengers began to titter and Roland glanced at me and scowled.

Now a tug of war began. The ferryman placed all his weight on the pole and heaved. The donkey, quick to grasp what was going on, stretched his neck forward enough to allow the raft a little movement and then, just when he seemed about to be hauled to his feet, lifted his head and strained back again with grunts of unbridled satisfaction.

Egged on by his passengers, even the ferryman began to see the absurdity of the situation after a while. A stupid grin slid across his face. Only Roland remained unmoved by this farcical turn of events. Eventually he rounded on the ferryman for a cruel and unfeeling scoundrel and demanded that he put back into the bank again. His remonstrations went unheeded, however, for at that moment there was a breathless cry and the figure of a man appeared, running along the bank and waving his arms. Some distance behind, also waving and yelling, pelted another two figures.

Drawing level with us, the leading figure dashed past the donkey and leapt for the raft. It rocked wildly as he landed, lurched forward and finally, aided by an opportune thrust of the pole, slid out into the river to loud applause, towing behind it an astonished Nebuchadnezzar who had been dragged off the bank and into the deep, fast-flowing water before he knew what had happened to him. Now his disembodied head bucked furiously at the end of the rope, gasping and snorting as Roland, down on his knees again at the edge of the raft, shouted encouragements.

The crossing was brief. We disembarked at a point some distance downstream where the bank shelved enough for the donkey to gain a foothold and scramble out. Only then did I glance again at the other passengers and realise with a start that I recognised the latecomer. It was the wheelwright. He appeared thoroughly dishevelled from his exertions and more than a little agitated but he had not yet taken note of us. He seemed more preoccupied at present with the opposite bank where the remaining two figures had come to a halt, still yelling and waving in evident frustration.

After a while he turned to the ferryman, pressed a coin into his hand and told him, in a low voice, to make the crossing his last. A sly look crept into the ferryman's eye as he glanced first at the coin, then at the far bank. He gave a conspiratorial wink to the wheelwright, pocketed the coin, stepped back onto the raft and began poling slowly upstream again.

Nebuchadnezzar was now rolling ecstatically in the long grass on the bank. Roland looked on dotingly. The wheelwright stepped onto the path, caught sight of him and stopped as if struck by an arrow. Roland glanced up and their eyes locked in mutual amazement.

It was the first time I had ever seen Roland lost for words. The wheelwright seemed equally tongue-tied and after a moment's silence I intervened with the first thing that came into my head.

'You got away . . .'

An uncomfortable look crossed the wheelwright's face as he glanced over his shoulder.

'From the reeve, I mean,' I added hastily, 'from the village.'

He eyed me for a moment, then said gruffly: 'That I did.'

The sounds of an altercation now drifted downstream. Once again the wheelwright glanced anxiously towards the far bank where three small figures stood gesticulating by the raft. At length one of them could be seen to shoulder the pole and stride off in the direction of a nearby cottage, leaving the other two to make their way slowly, and in obvious disgust, back the way they had come.

The wheelwright's taciturn expression lifted minutely and he turned to us again.

'What brings you here, then?'

Roland remained silent and turned away with an ill-humoured look when I tried to catch his eye. Ignoring him, I began to relate to the wheelwright what had taken place in the village.

We left the riverbank on a rough road which ran straight ahead through pleasant, flat grazing land, broken here and there by small woods and stands of poplars. We walked slowly in the gathering dusk, keeping a little ahead of Roland, who had made it clear that for the time being he preferred the company of Nebuchadnezzar. Now he was leading him by the halter as he delivered a monologue of inane endearments.

Apart from an appreciative grunt at the fate of the reeve's son, the wheelwright listened with little reaction to my story. When I reached the tragic conclusion of events at the manor he merely nodded, as if to say that that was no more or less than he had expected. I asked him, in turn, how he had fared since leaving the village but he shrugged evasively, then gestured ahead to where a cluster of lights were now springing up in the dusk.

'The town, see?'

I enquired if he knew it and he paused, as if wondering whether he might incriminate himself with his reply, then nodded.

'What takes you there?' I regretted the question as he looked at me sharply and said: 'Could ask the same of you . . .'

I mumbled something about the need to find work and was relieved that he did not pursue the subject. We walked on in silence and I wondered what it was that marked his path through the prevailing disorder, this stocky craftsman with his dour, suppressed energy and his unstated business. I thought of the map, now in the satchel on Roland's shoulder, and was suddenly grateful for it.

As we drew closer, the town walls began to loom before us. Above them, an untidy pyramid of lights outlined the shape of a low hill and only now did I begin to grasp the enormity of the place. There were more lights here on this one night, climbing into the darkness, than I had seen in twenty Christmases. The walls themselves were as tall as three laden haycarts, maybe four, and there was no telling how far they stretched away to left and right. A shadowy jumble of roofs became visible beyond them and each one seemed as big as the roof of the manor. The arched and lamp-lit gateway, towards which the road now led us directly, gaped like the waiting maw of some gigantic stone beast. In a moment of panic, I asked the wheelwright if he knew where we might be safe for the night and sensed his amusement as he grunted that he would show us a place.

Satisfying the scrutiny of the gatekeeper and his two beefy attendants, we paid our toll and passed through the archway into a narrow street. Although the way led us gently upwards, my first impression was that we had descended into purgatory: the houses teetered out to meet one another over our heads, leaving only a ribbon of sky, as if seen from the bed of a deep canyon. Lantern light spilled from shuttered windows to create weird, distorted shadows on the uneven surface at our feet; the cacophony of a nearby alehouse bounced eerily back and forth between the buildings and we were enveloped by a nauseating stench, as if the whole place had been doused with a hell-broth of rotten vegetables, fish guts and dead cats.

The acute sense of confinement made me cast anxious glances over my shoulder towards the gate, hoping for a glimpse of the open country beyond. But the walls still met the sky and I walked on at the wheelwright's heels, wondering how people could inhabit such a dungeon of a place. Behind us, Roland was still muttering softly to Nebuchadnezzar – more now, I suspected, for his own reassurance than for that of his charge.

At the top of the street we emerged briefly into an open space where the ground levelled and a number of other streets and alleys converged, then we were plunging down again through a rat-run even narrower and darker than the one we had just climbed. After a short, steep descent we were disgorged into another, much larger space which I guessed must be the market-place.

In the centre stood a worn stone buttercross with the melancholy silhouette of an empty stocks before it. On three sides were houses and other buildings of various description. On the fourth, directly across from us, rose the town wall. Sprouting higgledy-piggledy along its foot was a miserable encampment of shelters, whose ragged occupants sat lethargically around a small constellation of cooking fires. I was struck by the incongruity of the scene in such otherwise solid and prosperous seeming, if malodorous, surroundings.

'Beggars?' I asked.

The wheelwright shook his head. 'Country folk – fleein' the sickness. Got in afore they shut the gates. Not that it did 'em much good.'

'Pestilence? Here?' Somehow I had imagined that towns were inviolate – orderly places where money kept things in hand and the walls, as a last resort, deterred unwanted incursions.

He gave a short, humourless laugh. 'Took ten score. Near half of 'em.'

There were two inns facing each other across the market-place, with their signatures extended competitively at the end of long poles which drooped low over the heads of passers-by: the sign of the ox and the sign of the woolsack. The wheelwright made for the sign of the woolsack and went within to enquire about lodgings.

He came out again shortly, nodding, and indicated that we would find the stables at the rear of the building. Roland, who had still not said a word since leaving the river, led Nebuchadnezzar away and disappeared down the darkened side-alley. I hovered for a moment, wondering whether I should remain with the wheelwright, but Roland's mood was beginning to unsettle me and the urge to clear the air quickly overcame any inclination for courtesy towards our guide.

I followed Roland into the stable yard and stood for a while as he secured the donkey in its stall, spread out some hay and talked to it gently, stroking its head. He was not aware of my presence and as I watched his soft gestures and heard the tenderness in his voice, I found myself having to admit that here was a genuine affection. It startled me to realise, moreover, that I was even slightly jealous.

After some while he put his arms around the donkey's neck and pressed his head against its coat, resting there, quite still. I walked towards him and as he looked up I could see, by the light of the single lantern hanging on the inn wall, that his cheeks were damp with tears.

I muttered an awkward apology for having appeared unconcerned about Nebuchadnezzar's welfare. Roland shook his head wearily and ran the back of his hand across his eyes. 'It's not that, Creb . . .'

'What is it, then?'

Again he shook his head and shrugged. 'Nothing . . . everything.'

He turned away and began to stroke the donkey again in a distracted way, his shoulders slumped miserably in the shadows of the stall. I took his arm and suggested that he would feel better once we had eaten. He looked up and smiled weakly.

The public room was long and low and acrid with woodsmoke which, due to some architectural imperfection, seemed reluctant to leave the fireplace by the chimney. There were fifteen or twenty customers seated on stools at a scattering of rough tables. Some were merely drinking, others using hunks of bread to scoop a greasy-looking stew into their mouths. A couple of serving women bustled about between the tables, their banter rising above the general hum of conversation.

The wheelwright, who had taken the last place at a table near the fire, acknowledged us as we came in, pointed vaguely at some free places on the other side of the room, then returned his attention to the stew. I took it that he had had enough of our company for the day.

We made for the table, ordering food and ale on the way, and sat down gratefully. The warmth of congregated bodies and the buzz of voices was soporific and as we waited for the food to appear I glanced around the room, gradually aware of an unfamiliar, self-contained pleasure in the anonymous company of strangers. Even if the filtered snatches of conversation were predominantly concerned with the pestilence and its aftermath, the voices, and even the expressions, seemed somehow mellowed by smoke, lamplight and ale. An inn, I thought, was a fine place to be.

The ale, which arrived in advance of the food, was also fine. It drenched the palate with a sharp sweetness that contrasted delectably with the vile brew to which we were accustomed at home. Our ale tasted and smelt like fermented badgers' urine and was so weak that a decent state of intoxication was beyond the reach of anyone without a cauldron for a bladder. I sat and sipped and savoured, reflecting on all the years during which I had taken that noxious brock's piss for the culmination of the brewer's art.

Our food arrived at the same time as two newcomers, who took their places opposite us. One was a middle-aged man in a dirty hooded travelling cloak. He removed the hood as he sat down, revealing the face of a minor brigand: sallow complexion, long unshaven jaw, lank grey hair and an artful expression. The other, a young woman of about my own age, was similar in colouring but with features so much finer and softer that it was hard to make the obvious assumption that she was his daughter. She glanced across tentatively with large, intelligent brown eyes but did not return my smile.

The fellow, meanwhile, was blatantly assessing us both. At length he leaned forward and in a voice larded with bonhomie, said: 'Travellers, eh? Expect you young fighting-cocks have a tale to tell. Very partial to a story m'self – while I'm eating. Care to oblige?'

Roland and I glanced at one another and I caught a flicker of something close to disgust in the girl's eye. She coloured and looked away. But her father, if that was what he was, was not to be deterred. 'Come on, my lads. Travellers' tales at the inn.

Companionship of the road . . .' His eyes widened as if something marvellous had just occurred to him. 'Tell you what, here's a bargain. Tell me yours, I'll tell you mine.'

Roland's expression suggested that the last thing on earth he wished was to hear this fellow's story, but the prospect of a prolonged and awkward silence seemed worse, so I began to relate, very selectively, the events so far. Once or twice I caught the girl's eye and I had the sense that there was something she wished to say to me, but could not.

When I had finished, the brigand wiped his chin on his sleeve and said: 'An adventure, my lads. An adventure indeed. Now, a bargain's a bargain . . .'

Whether it was because of the ale which had begun to creep round my brain making me feel flushed and light-headed, or because there was some genuine wit there, I found myself becoming increasingly entertained by his story. It was almost all invented, I was certain, but that did not seem to matter. I laughed heartily and as he drew to the conclusion of one anecdote, encouraged him to follow it with another. I noticed that Roland was drinking little and looking less than enthusiastic, but put it down to his earlier mood, for which I seemed suddenly to have lost sympathy. He had only himself to blame if he could not enter into the spirit of things. After some time, he rose and went off to relieve himself. The brigand took advantage of his departure to draw breath and nonchalantly scan the room. At some point he appeared to catch someone's eye and a brief, wordless exchange took place. He turned to the girl and muttered peremptorily. A wounded look came into her eye, but she rose obediently and I watched her make her way between the tables, thinking muzzily how much more pleasant it would be if she stayed. Then I caught sight of Roland re-entering the room. As he passed by the fire the wheelwright beckoned to him and spoke briefly, glancing in our direction.

'Friend of yours?' asked our new companion sharply.

I shrugged. 'Met him at the river.'

'Going on with him, then?'

For a moment I could not remember. 'He has some business here, I think.'

'I like a man who does business.'

Roland returned, glanced at the empty stool and sat down again.

I asked him what the wheelwright had wanted. He gave me an uncomfortable look and muttered something about Nebuchadnezzar and the stable.

'Nebuchadnezzar?' The brigand gave a loud guffaw. 'Nebuchadnezzar?'

'My donkey,' said Roland.

'Necky . . . Nebby . . .' My inability to wrap my tongue around the ludicrous word pitched me into a sudden paroxysm of laughter.

I could not understand why Roland looked so disapproving, but the brigand grinned with me and began to tap out a rhythm on the table-top. 'Nebuchadnezzar – King of the Jews – sold his wife – for a pair of shoes – when the shoes – began to wear – Nebuchadnezzar began to . . . can't remember what. Good rhyme, that, though. Good name, too.'

Roland appeared startled. He raised his cup hurriedly, took a gulp of ale and pushed it aside. Then he rose to his feet with a terse sideways glance and said: 'Pardon me. I'm very tired.'

'G'night then,' I said.

'You too, Creb.' He took my arm. 'Come along. It's been a long day.'

I pushed him away. 'I'm fine. I'll come later.'

He hesitated, then shrugged and left the room.

'Tell you something,' I said, tilting back on my stool and waving expansively.

'What's that?' said the brigand.

'We're going to get rich, y'know.'

'Are you, now? And how's that, young Creb?'

'We're going to be . . .' The word suddenly eluded me. 'Lead into gold . . . you know . . .'

'Aha!' The brigand cocked his head on one side. 'Alchemists, eh?'

I nodded.

'Lead into gold?'

'Lead into gold.'

He looked at me for a long time, then leant forward conspiratorially. 'Or could it be something else?'

'What?'

'A matter of the spirit perhaps . . . the soul . . . the inner being . . . the flame that burns bright in the night?'

I looked at him blankly and he sat back with a chuckle. 'Lead into gold it is, then. And good luck, I say.'

Somewhere in the soupy interior of my skull, a small voice told me that I was being mocked.

Some time later I made my way unsteadily up the stairs and into a darkened attic room where mattresses were arranged all the way around the walls. A number of them were already occupied and since I could see none too clearly by now, it was only on the third attempt that I picked the right bedfellow and slumped down beside Roland.

'You're drunk,' he hissed, as the chorus of startled grunts and curses died away.

'I'm not.' I closed my eyes and turned an enormous, oily somersault, then another. I opened them again and sat bolt upright.

'Yes, you are. We should never have been with that fellow in the first place.'

'Says who?'

'The wheelwright.'

'Why?'

He leant across and began to whisper something, but at that moment I knew I was going to vomit. I fled downstairs and into the market-place where a large yellow dog growled and fixed me with an unsympathetic stare as I retched onto the pile of refuse it had been inspecting.

I returned to the attic again, feeling a little better, and tiptoed across to the mattress.

'I have to tell you this, Creb,' whispered Roland urgently as I lay down again. 'Are you listening?'

'Yes.'

'The wheelwright said we shouldn't be with that fellow because he's a Jew – and you should never trust a Jew. They're blasphemers, fornicators, poisoners. . . That's what the wheelwright said, Creb.'

'Are they?' I said disinterestedly. 'He seemed all right to me. Good stories, too. I liked him.' My head had begun to pound and all I wanted now was to sleep.

'Well, I didn't. I didn't like the way he looked at me.' He paused. 'And for that matter, Creb, it wasn't him you liked. It was the girl.'

Nine

was awoken next morning by the none-too-gentle prod of a foot. I opened my eyes to discover that during the night a forge had been set up inside my head. Its proprietor hammered with berserk fury at the walls, the floor, the ceiling and, occasionally, the anvil from which sparks flew like red-hot pins.

Roland stood with hands on hips at the end of the mattress, fixed me with an unsympathetic stare and commanded me to get up. He had something to tell me.

As I raised myself first to my knees and then, very slowly, to my feet, I discovered also that I was itching quite horribly, all over. There was no softening of Roland's expression, not even a glimmer of amusement, as I raked myself with my fingernails and gasped at the relief which, though momentary, was beyond description.

We left the room and made our way downstairs. Vague cameos of the previous evening began to flit through the reverberating gloom in my skull: a disreputable-looking Jew who had made me laugh while he poisoned my ale; his silent, pretty daughter with weariness and pain in her eyes; something dimly troubling to do with lead and gold and a flame in the night . . . and then, dear God, Roland himself. The scene in the stables came back to me all at once and with it, my good intentions of lifting his spirits, then how the Jew and the ale had intervened and how distracted and short-tempered Roland had become as the evening had progressed.

I began to apologise, but he cut me short: 'I've been out. Making enquiries. The castle's a two-day journey from here.'

I congratulated him and tried to smile but the sunlight and noise in the market-place, where the day's business had already begun, seemed to intensify the ill effects of any muscular activity, however insignificant. There was a watering trough outside the inn. I sluiced my head and neck repeatedly with greenish, scummy water.

The appetising smell of fresh bread thickened the air and it came to me that I should put something in my stomach. Roland grudgingly admitted that he also was hungry and we threaded our way through the stalls towards the source of the smell – a clay oven, squatting on the ground like a large, pale molehill. Beside it, a fat woman with a disapproving expression stood guard over a battered tray of small, flat loaves.

We made our selection and I fished inside my breeches for the now familiar knot of coins. The shirt-tail hung loose against my leg. As casually as I could, I asked Roland if he had taken charge of the money the previous evening. He paused with a loaf halfway to his mouth, glanced at me sharply and shook his head.

I could feel the blood rushing to my cheeks as I searched myself once again, more thoroughly, and Roland stood by in grim silence. Then another memory stirred: of ordering more ale for the Jew and myself; of reaching beneath the table in a fuddled attempt to take my breeches down so I could re-knot the money into my shirt-tail; of the prurient gleam that had suddenly entered the Jew's eye; and of my hasty abandoning of the attempt.

I hung my head and patted my jerkin pocket. 'I put it in here. Must have fallen out.'

For a moment he stared at me with a coldness I had not seen before, a look which instantly deterred me from following him as he turned and walked back to the inn. I waited there wretchedly until he returned, a short time later, holding one small coin between thumb and forefinger.

'Under the mattress,' he muttered disgustedly, proffering the coin to the baker-woman who had stood watching our drama unfold without changing expression. Now she shook her head and silently extended a large, sweaty palm. With a grunt of exasperation, Roland dropped his loaf into the outstretched hand and walked away. My heart dropped to my boots as I passed her back the rest of the loaves. I turned to go, but she tapped me on the shoulder.

'Drunk, was it – at the inn?'

I nodded, wondering what I must look like.

'You'll not do it again.' She winked and handed me back one loaf.

I nodded again, grateful and contrite, then made my way back through the market to where Roland had propped himself against the water-trough. I offered him the loaf. After a moment he took it and turned it in his hands, inspecting it minutely as if it were the last morsel he would ever eat. Stricken with shame, I began again to apologise but this time he merely shrugged, then broke the loaf in two and passed me half.

For some time he chewed in silence, staring blankly over the heads of the throng before us. Then, without looking at me, he said: 'No money, Creb, means no food. No food means no journey. No castle. No stone.' He paused. 'What do you suggest we do?'

There was as much disappointment in his voice as anger and I remembered with a pang how I had felt after the episode of his aunt's bath. Unable to think clearly, I squirmed in miserable silence, then jumped as someone tapped me lightly on the shoulder. I turned to see the Jew's daughter standing behind us. She seemed neither confident nor timid but there was a dull resolve in her eyes which replaced the vulnerability of the previous evening, leaving her curiously less attractive, I thought.

'Beg pardon, but I was by the baker's. I couldn't help hearing . . .' She paused, then continued hastily: 'If you need money, I can get it.'

Roland eyed her with immediate suspicion. 'How?'

'I'll tell you – but you'll have to take me with you.'

'Where?' I asked.

'Wherever you're going.'

'Why?' Roland asked sharply.

'Because . . .' she began to chew her lip, 'I have to get away from him.'

'Your father?'

She shook her head wearily. 'Uncle.'

'Why,' Roland persisted, 'must you get away from him?' I could not understand the hostility in his voice. He was normally so attentive, so eager to engage with newcomers.

She seemed scarcely to notice, hesitating for only a moment before replying: 'Because he makes me whore for him.' It was spoken softly, without emotion – as if she had rehearsed it, knowing that this confidence was what she would have to trade for her freedom.

I made way for her on the edge of the trough and as she sat down, enquired why she had not left him before. She glanced at the town walls and explained that she had considered it too dangerous to be alone on the road. She added that she had not yet come across anyone she felt she could trust.

'You trust us?' I intended to sound light-hearted but she merely looked at me, very directly, and said as if it were quite incontrovertible: 'You're honest.' She glanced at Roland. 'Him too.'

Roland's eyebrows flickered and his voice seemed to lose a little of its edge as he enquired where she planned to obtain the money.

'From him.' She gestured towards the inn.

'Will he give it to you?'

'No. But I . . . earned it for him. And I've not seen a penny of it . . . nor wanted to, till now.' A certain calculation entered her voice. 'But it won't be hard. He's a sodomite, my uncle.' Roland winced. 'Keeps the money in a belt, under his shirt.' She looked at me again. 'You could do it. Some quiet place. He's not strong. A coward, too.'

Roland began to say something, then changed his mind and frowned.

My own misgivings must have been equally apparent for she waited a moment, then said: 'What else will you do for money?'

I shrugged. 'But . . . wouldn't he make trouble?'

'Who cares for a Jew?'

'You are one.' Roland said it without accusation, a mere statement of fact, but for the first time her voice became animated and I noticed her hands clench. 'We're not all like him.'

Roland's eyes lost their focus again. For a while he remained silent, gripped by some inner conflict which caused his brow to furrow deeply, but at length he looked across at me and, to my considerable surprise, gave the faintest nod.

'What do we do, then?' I asked.

She held up her hand. 'I need your word first – that I can come with you.'

'You have it,' I said.

'And yours?' She gave Roland a penetrating look and he nodded.

Something flitted across her face – hope, relief, I could not tell what – and for a moment the softness of the previous evening returned.

The Jew was at my heels as we left the inn. I could almost feel his breath on my neck.

On the opposite side of the market-place an alley led past the stables at the sign of the ox, dipping down towards a warren of smaller, poorer dwellings contained by an angle of the town wall at the foot of the hill. We had explored the place earlier and found an abandoned shack which would suit our purpose.

I walked fast, thankful that the alley was too narrow for us to go abreast; thankful also that he seemed too aroused to speak. I had seen the bonhomie and humour evaporate, driven out by the rush of blood to his face, the sudden glazing of his eyes, as I had made my faltering proposition in the public room of the inn. He had muttered something thickly and followed me outside, leaving unfinished ale on the table.

Shortly the alley petered out and we found ourselves picking our way through what was little more than a dense hutment of low, single-room hovels and ramshackle animal enclosures, foul with the stench of ordure littering the ground between them. An all too appropriate setting for this endeavour, I thought, rounding the corner of a pigsty which I recalled from our reconnaissance.

But the dilapidated well I had expected to see was not there. I stopped and looked about me, watched impassively by two raga-muffin children and a gaggle of slovenly ducks.

'Where are we going?' He sounded impatient and was fumbling with one hand inside his cloak.

'Somewhere quiet.' I tried to make it sound as if this was something I did all the time, yet a part of my mind seemed unable to accept that I was doing it at all. It left me with the disconcerting sensation of being scarcely in contact with the ground beneath me.

We set off again, followed by the children. Did I recognise that line of grubby washing . . . that old man asleep in his doorway . . .? My palms were beginning to sweat. The town wall was visible at every turn and the shack, where Roland and the girl were now wait-ing, was hard against it. It could not be that difficult to find . . .

His hand came down heavily on my shoulder. 'Can't wait. In here!'

He began to push me backwards into the shadows of a tunnel-like space between a dwelling and some sort of a small cowhouse. Over his shoulder I could see the children watching, intently now. I caught the eye of one of them and they scuttled off, squealing and giggling.

The Jew was panting, struggling with one hand to unclasp his belt as he pushed me deeper into the gloom beneath the overhanging thatch. For a few indecisive moments I allowed myself to be manhandled. Then, almost of its own accord, my knee came up and caught him between the legs. He was already sinking to the ground, eyes dilated in agony, as my fist struck the side of his face and he grunted a second time and fell sideways.

I cannot remember precisely what happened next, but some indefinable time later I was standing unsteadily in the shadows, looking down at the Jew who lay silent and inert at my feet. His face was bloodied and the pale skin of his torso was visible beneath his torn shirt. I walked slowly into the daylight, shocked and trembling.

It was not until I had gone several yards that I remembered why I had come here in the first place. I retraced my steps, glancing apprehensively around me. Then, feeling deeply ashamed, I knelt in the gloom and unbuckled the money belt. He grunted as I heaved it out from underneath him, but did not move. I stood up and fastened it round my waist where it hung under my jerkin, loose and heavy.

Clinking like Iscariot, I set off to find Roland and the girl.

We had left hurriedly and were now some way off across the plain. Behind us the town perched indistinctly on its eminence with the fresh scar of the burial pits at its foot. The weather had turned mild and we could feel a suggestion of the sun's warmth as it appeared between shifting clouds, which threw swathes of shadow across the meadows and woods around us.

Nebuchadnezzar, refreshed by a night in the stables, had entered into the spirit of our hasty departure and Roland was having some trouble in keeping him from a trot. The girl and I had been too breathless to talk and too preoccupied, in any case, with the possibility of pursuit. For some time we had continued to glance over our shoulders, but no riders had burst from the town gate and now we had all slowed to a more even pace.

Her name was Ellen, she told us, and she could cook and stitch. They had lived, she and her family, in some out-of-the-way town in the east of the country – the descendants of a line of itinerant tailors who had crossed the sea and finally come to settle in the place of her birth. There had been a handful of other Jews in the town, craftsmen and small merchants for the most part, but they were an unobtrusive little community of whom the local people took scant notice. Miraculously, this disinterest had persisted even when the oppression of the Jews had been at its most virulent: whilst their brethren elsewhere were being vilified, robbed, assaulted and ultimately expelled, Ellen's forebears went unmolested. Three generations later, the family continued to lead a modest but cheerful existence, discreetly observing the small rituals that distinguished them from their neighbours, and troubling no one.

Then the pestilence had come, mysteriously striking at the poorer, more crowded part of the town first. Within a few days the rumour had begun to circulate that the Jews, most of whom had so far been spared, were to blame – that they had introduced the sickness deliberately to the town's water supply. The Jews had responded swiftly with a delegation to the mayor, led by the apothecary who had already begun to employ a foul-smelling potion of his own devising to some effect amongst the smitten. Fearful of losing their only apparent prospect of salvation, the mayor and his burgesses had threatened to fine anyone heard advancing the scurrilous theory and, for a while, the frightened and bewildered population had taken note and the rumours had died away.

But the pestilence had not. With each new death, the people of the town grew more fearful, less comprehending, more reckless. The rumours started up again and then, one day, a band of strange men and women wandered into the town speaking a foreign tongue.

Their leaders carried tattered banners of purple velvet and cloth of gold. The followers were cowled with sombre cloaks upon which a scarlet cross stood out like an intimation of hellfire. They stopped in the market-place where one of their number stepped forward and explained in halting, guttural tones that they were members of the Brotherhood of the Flagellants who had set out on a journey along the coast of their own country, but their ship had been blown off course by the Almighty in order that they should save the souls of this poor land. Whereupon the men removed their cloaks and began to chant as the women set about them with scourges and flails until their backs were raw and they were fainting to the ground. They were driving out impurities, the spokesman explained, and the people of the town would do well to follow their example.

Their flagellation over, the foreigners collected alms and moved on and the crowd made straight for the church. An hour or so later, transformed by now into a hysterical mob, it emerged again, led by a woman brandishing a leathery scrap she claimed to be the foreskin of an infant, retrieved from a well. So sudden and unexpected was this grotesque turn of events that the Jews were taken completely by surprise. Men, women and children, they were rounded up and taken beyond the town walls where they were herded like cattle into a wooden barn – and there, burnt alive.

Her uncle, a draper, had that morning commandeered Ellen to accompany him as his assistant to a nearby manor, whose lady had sent word that she wished to inspect samples of his cloth but dared not enter the town. They had returned in the evening to be confronted by a smouldering heap of timbers and charred, vaguely anthropomorphic bundles – all that remained of their family and friends. They had turned tail and fled into the forest.

She told the story as if it had happened to someone else and although she must have suffered greatly – by the unspeakable fate of her family, the hardship of her subsequent wanderings with her uncle and the misery of being made to earn her keep – she seemed almost determined not to arouse sympathy.

It took her some time to tell the whole tale. I had ample opportunity to study her as we walked along at Nebuchadnezzar's side and to wonder about the contradiction in what I saw. Her face was lovely, with a soft, slightly olive skin, fine cheekbones and large brown eyes beneath dark lashes. Slim wrists, delicate hands and small feet suggested a slenderness beneath her enveloping cloak. But radiance and grace were absent. She had not yet smiled, nor really shown any other trace of animation beyond a flicker in the eye, and her movements betrayed an inner weariness. She seemed, not surprisingly, to be utterly at odds with herself.

At length she finished her story. The ensuing silence was strangely oppressive, an emptiness charged with unspoken suffering, and I found there was something within me that wanted to reach out to her, to fill the void at least, even if I could not share her thoughts and feelings. As the road meandered towards the low hills at the opposite edge of the plain, I began to tell her a little of what had befallen us. She listened indifferently and did not seem to notice when, after a while, Roland gave me a sharp glance and intervened to take up the story himself, carefully omitting any mention of our destination or the reason for it. I felt irritated by his obvious lack of trust but said nothing and let him continue. Listening with only half an ear, I found myself thinking how important stories had become in the last few days. Without possessions or any of the other marks of status, it seemed that our story was the only thing that substantiated us to our fellow travellers. It had become almost a kind of map, lending coherence to an otherwise unintelligible terrain . . .

In the late afternoon we left the road and climbed slowly into the hills on a rough track, following the directions Roland had been given. Towards sunset we came upon a deserted barn on the edge of a high meadow. There was a village in the valley behind us and as we foraged for wood in the dusk we could see the lights springing up below. I felt glad to be away from people again.

We made a fire on the earthen floor of the barn. Quietly and efficiently, Ellen took charge of the cooking. She skinned and gutted the rabbit we had bought earlier on, cut it into small portions and placed them with chopped vegetables in the cooking pot. Her movements were deft and precise. A little frown of concentration reminded me of my mother.

We sat by the fire, savouring the smell from the pot. It may just have been the firelight, it may have been the pleasure of the small domestic routine, but it seemed then that she was almost content. She knelt forward from time to time to stir the pot, then sat back again with a nod of satisfaction.

Roland had been pensive and distant throughout the early part of the evening but the food appeared to bring him back amongst us. As he sucked the meat from the last rabbit bone and wiped his mouth with the back of his hand, he nodded appreciatively and congratulated Ellen on her cooking.

To my delight, she responded with a smile and her face became everything I had imagined it could be. Roland blushed and it occurred to me that he might never have experienced the company of a young woman before. At any rate, the brief exchange seemed to release something in both of them. The firelit interior of the barn suddenly became a warmer, more companionable place and we talked on into the evening under the watchful gaze of three owlets, whose eyes blinked like a distant constellation in the darkness of a crevice, high above us, as they awaited their mother's homecoming.

But we had all three fallen asleep around the fire, long before her silent return.

Ten

The castle stood in the lee of a crescent of wooded hills. A garlanded boar's head fluttering over the gatehouse left us in no doubt that we had found the right place.

I had never seen a castle before. It seemed inconceivable that anyone could have piled stone upon stone, day after day, to build something so immense. The ramparts were pale, almost honey-coloured, glowing in the sunlight against their dark backdrop and towering benignly above the village sprawled at their feet. But the turrets and bastions rising at intervals along their length served as a reminder of their real purpose, and peering over the top, as if to keep an eye on us as we approached, were the windowed and embrasured upper storeys of the keep.

We made our way towards the village in good spirits. Over the last two days Ellen seemed to have undergone some slow, cautious process of emergence, like an animal at the end of its hibernation. The further we distanced ourselves from the town, the more she seemed to step into her proper rhythm. Gradually a sense of lightness, at times even gaiety, crept into her looks and movements. And in a sudden flash of the eye, an impulsive toss of the head, I was beginning to glimpse a wilfulness which had lain well concealed behind the lacklustre mien we had first encountered. If I had been struck at first by her prettiness, I was now becoming equally struck by her spontaneity.

Roland also seemed at last to have shaken off the fragility of mood that had dogged him these past few days and in a quiet moment the previous morning, while Ellen had still been asleep, we had resolved our differences with a certain amount of solemn apology, followed by sheepish grins, then a bout of jolly backslapping and robust protestations of friendship. But he was not entirely unchanged; where there had formerly been uncontained exuberance and curiosity, there was now a trace of reserve, as if he no longer permitted his words, looks and gestures to express the sum of his thoughts or feelings. It confused me, this feeling that he had drawn a veil, however flimsy, around his core. I found myself troubled by momentary stirrings of regret, resentment and even guilt that I could only half explain. But I kept them to myself, and as I watched Ellen beginning to warm to him, I had to admit that this new reserve rendered him more intriguing than ever. He, meanwhile, seemed to have become less flustered by her attention. On the previous day, she had asked him why he rode rather than walked, and he had readily explained about his fits and his shortness of wind. From then on they had struck up an easy, bantering rapport and a little while later, as we paused at the side of the path to eat our bread and cheese, Roland embarked on a lengthy explanation of our journey and its mission. From time to time, as he spoke, I would surreptitiously seek Ellen's eye and savour the quickened heartbeat as she did not immediately look away.

As we reached the outskirts of the village we overtook an old woman driving a small flock of geese. We stopped and asked her if she knew of a gibbet. She gave us a long, rheumy look before replying that it depended which one we meant. There was one at a crossroads about half a day's walk in that direction – she pointed east towards distant grassy uplands – but it had not been used for months, not since the pestilence came; there was another, some miles into the forest beyond the castle, where yesterday – or had it been the day before? – the lord's men had strung up a couple of robbers, but it was not so much a gibbet as an old hanging-tree.

We thanked her and walked on a little distance before stopping by a stream to consider our alternatives. Roland brought out the map and spread it on the ground and all three of us knelt to look at it, Ellen's shoulder against mine.

The road ran straight up the page, about a thumb's length, from the castle to the gibbet, where it broke briefly on both sides. Although no other road was indicated to left or right, the symmetry of the break suggested that it was deliberate and that we should therefore interpret it as a crossroads. But there were also trees marked all around it, and we found ourselves faced with the confusing possibility that there was both a crossroads by the hanging-tree, and a clump of trees by the gibbet at the crossroads.

Although Roland favoured the gibbet, while I was for the hanging-tree, he appeared to lack some of his usual conviction and I knew quite well that my own preference had no very sound basis. Ellen listened to us arguing for some time without venturing any opinion herself, then stood up abruptly and walked over to Nebuchadnezzar, who was browsing at the side of the path. She leant against him and stared thoughtfully into the distance, then turned with a look of exasperation and said: 'You haven't understood, have you? Whichever one we choose, we won't know whether it's right until we reach the next place – the hill. And how will we recognise that? There must be hundreds.'

Roland looked up and nodded. He seemed impressed by this observation, as was I. In all the thrill of discovering the meaning of the map, neither of us had stopped to consider whether it would function as the description of an actual route. Now, as we began to reflect on the implications of Ellen's remark, I found myself becoming intrigued again by this idea that a wilderness of hills, woods, rivers and plains could somehow be given sense on a flat, lifeless piece of parchment; and swiftly depressed by the conclusion that it could be no more than a fairy-tale, as we went on to grapple with the apparent impossibility of identifying one hill – a mere hump on the page – from its neighbours, or indeed one church steeple, one smithy, or one of anything else, especially when there were no indications of direction or distance. In the end it was Ellen again who suggested that perhaps the map was drawn to some sort of scale. The irregularity of the distances between the main features along the road seemed too pronounced to have been accidental, and it would be a simple matter to note the time, at a steady walking pace, between one feature and the next.

This time a peevish look entered Roland's eye, but it was swiftly replaced by a glimmer of satisfaction as he pointed out that that still did not resolve the immediate dilemma of the crossroads.

Ellen glared at him for a second, then shrugged and smiled.

'We could leave it to fate . . .' She fished in her cloak and produced a coin, holding it up to the sunlight between her fingers. I could not help noticing how pearl-like her delicate pink nails were.

Roland eyed the little disc for some time before acknowledging the absence of an alternative with a reluctant grin. Ellen flipped the coin. It spun, glinting, and settled on the grass, favouring the hanging-tree.

The way through the forest was clearly marked. The undergrowth had recently been hacked back to a distance of three or four paces on either side of the path to create a mottled green canyon, meandering through the trees.

We first saw them at some distance, directly ahead of us. The two bodies bore little semblance of humanity; a pair of puppets dangling from the outstretched branch of a large oak. Turning slowly in the breeze, they might have been mere woodland scarecrows. It was a sight at first to arouse the curiosity, but as we drew closer a sense of revulsion began to take hold. The monk had been one thing – his life had slipped softly from him, coaxed away by the hunger and cold, but these two had clung on till the end, twitching and wriggling as their tongues swelled and their faces darkened.

When we were twenty yards off, Nebuchadnezzar snickered and halted. Roland dismounted to try to lead him forward, but the donkey refused and had to be tugged from the path and led into the trees on a wide detour around the oak. Ellen made a little noise of disgust and followed, her gaze fixed on the ground. I remained where I was, momentarily rooted in morbid fascination. I had attended so much death in the past weeks, had even caused it, and now I felt a curious duty to observe, to be reminded of, and give thanks for, my own vitality.

The breeze intensified, a branch creaked and one of the corpses turned leadenly on the end of its rope to face me. It was the wheelwright, eyes protuberant with outrage. I stared at him for some while, vaguely wondering how he had come to be there; conscious also that this final meeting caused me no great surprise. I was beginning to understand that chaos has its own subtle rhythm, where no beat is out of place.

'Creb!' Roland's voice was sharp with disapproval.

I passed beneath the oak and walked on a few paces to find them standing in a second, slightly narrower ride, which crossed our own at right angles.

'The crossroads!' he said with satisfaction.

I nodded, then glanced at the sky to gauge how long we had

been travelling. There seemed little point in mentioning the wheelwright.

Roland looked up too. 'About four hours,' he ventured.

'Nearer five,' said Ellen.

We consulted the map again, agreeing to settle on the half of one day as the duration of our journey so far. Ellen drew a thread from her cloak and measured the distance between the castle and the gibbet, then the distance between gibbet and hill.

'Half as far again,' she said. 'Three-quarters of a day's travelling, then.'

'If it works,' said Roland.

'If this was the right gibbet,' I added sourly. I felt suddenly irritated by the imprecision of our itinerary.

An hour later we stopped, still in the woods. We had been hoping to reach open country for the night, but as the shadows lengthened and the trees continued to march ahead of us into the dusk, I sensed Ellen becoming apprehensive and suggested a halt before it grew dark. A night in such evidently frequented forest as this was not ideal under any circumstances, but the prospect of a camp with a fire seemed preferable at least to the insecurity of being on the move in the darkness.

We left the path and made our place at the foot of an enormous beech which loomed above us like a petrified giant against the evening sky. There was reassurance in the moist, sweet smell of humus seeping through the dead leaves beneath us, and in the massive, gnarled roots all around us, shoulder-high as we sat. By the time the fire was going, and Nebuchadnezzar tethered a little beyond it as look-out, it seemed that Ellen's anxiety had left her.

I sat back, talking idly with Roland while Ellen prepared the meal. As the evening chill settled, Roland and I draped ourselves in the blankets Ellen had brought with her from the inn. But she, being closer to the fire, had removed her cloak and now, for the first time, I was able to see something of the body which had been visiting my imagination over the last few days.

She was wearing a long skirt and a bodice of the same coarse weave; not so coarse though that the fine curve of her haunches was not clearly visible as she squatted by the fire, the fabric pulled tight across her buttocks and thighs. The bodice was loose-fitting, long-sleeved and unadorned – a functional garment, graced nonetheless with a low sweep at the front which revealed her delicate shoulders and neck to startling effect in the firelight. A single unpolished stone – a mere pebble, it seemed – lay on a slim leather thong at her throat, accentuating the sheen of her taut pale skin. She leant across to add something to the pot and as the bodice fell forward I caught a soft, shadowed glimpse of surprisingly full breasts, momentarily unconstrained by the fabric.

She sat back again and turned towards us, catching my eye. She was half smiling, as if at some small inner amusement, and I wondered for a second whether she had revealed herself deliberately. Covering my confusion, I returned her smile and her lips parted generously in response. I looked for something coquettish in the large brown eyes, but could see only warmth and pleasure.

I leant back against the root behind me, intoxicated with the wonder of what was taking place. This was as different from anything I had experienced in the village as . . . as gold from lead. The fire in my loins, I knew. It was easily kindled and quenched. But here too was an unfamiliar flame in my heart, which conjured a sense of wellbeing so complete that I was overtaken by the sudden urge to communicate it to Roland. I turned to him and told him, with great sincerity, how much healthier he had begun to seem, and how glad I was for it.

He looked a little startled by this unprovoked effusion, but mumbled that he *had* been feeling a good deal stronger for a few days now. Since we had bought the donkey and he no longer needed to exert himself, he had gained colour. Now he chose to walk from time to time and paced himself when he did so. His slight physique seemed almost imperceptibly to be firming and there was a hint of robustness entering his features.

'I had my doubts at first,' he added with a grimace.

So had I, I thought, but said instead: 'You only needed to get used to it.'

He turned to me with that look I had come to recognise as the preface to one of his truths. Ellen caught it, too, and went still.

'No I did not, Creb. I needed to be nursed along.' He looked to Ellen. 'He's a good friend, you know. He's already saved my life once . . .' he paused 'and you . . . I think you'd do the same.'

Ellen flushed and returned her attention to the cooking.

I glanced sideways at him, touched by the return of the old

transparency, the glimpse of all the courage and generosity within that fragile, fine-boned frame. I laid my hand on his arm and squeezed, deeply regretting the inadequacy of the gesture.

I awoke to feel a fine, cold drizzle moistening my forehead. I could hear it pattering on the ground and sizzling faintly on the embers. In the darkness beyond, Nebuchadnezzar was moving about, snorting softly. And there was another sound, the one that had penetrated my sleep; an intermittent scuffing of dead leaves.

I lifted myself slowly onto one elbow and felt for my knife, as fear crept like an army of insects across my scalp and neck and down my spine. Beside me, Roland and Ellen still slept, cocooned in their blankets. I nudged them both awake, sensing their sleep-heavy eyes widen with concern as I put a finger to my lips and pointed into the darkness.

The scuffing continued, growing closer. More than one pair of feet, it seemed. They could not have heard our movements yet or they would have stopped. Only the fire revealed our whereabouts and that was too low to cast any real light, but the sounds that betrayed our visitors would give us away too, if we tried to move. The leaves were several inches deep here, where they had drifted amongst the roots. We could only stay where we were.

I indicated to the others to feign sleep and eased myself cautiously to the ground again, with one half-closed eye fixed on the fire. After a few moments the movements stopped. Nebuchad-nezzar neighed softly once, then fell silent. I longed to be able to see beyond the fire as we lay there, strung like longbows, listening to our own heartbeats.

The scuffing began again and shortly two figures became visible making their way around the fire, one on either side, bent low it appeared.

I waited until one of them was almost at my feet, then rose with a lung-searing yell to find myself towering over a small, wide-eyed boy no more than ten years old. There was scarcely time for the shock to register before his right arm jabbed and there was a brief, warm pain in my thigh. I lunged forward but he darted out of my way and would have escaped but for the root over which he tripped and pitched headlong into the leaves. Stepping across the root, I bent down and hauled him up by the collar, at the same time grabbing the waving knife-arm at the wrist and twisting until the weapon fell from his grip. Then, with the wriggling child held firmly at arm's length, I turned to Roland and Ellen.

Their assailant was not much older than mine and had fared no better.

They had him against the tree, one by each arm, where he hissed and spat like a wildcat, quite unimpressed by the cooking pot with which Ellen threatened to brain him. His knife, too, lay on the ground, a dull smear across the blade-tip. Roland and Ellen were both breathing heavily, but neither appeared to be wounded and it was clear that they were wondering what to do next.

'Hold him,' I said. My mouth was dry and my legs trembled.

Dragging my captive with me, I swept branches onto the embers and walked out beyond the fire to where Nebuchadnezzar stood quietly at tether. At his head was a middle-aged man, clumsily swathed in animal skins. He held one calming hand on the donkey's neck, the other extended before him, loosely hefting a knife. The branches caught, flaring behind me, and I could see him eyeing me coldly as I advanced.

I stopped a little short, endeavouring to keep the fear from my face as I struggled for an opening to this confrontation. The man remained still and silent, his features devoid of expression. It was only when the child began to squeak in protest that I realised I had been viciously twisting its collar, and inspiration mercifully intervened.

'This yours?' There was a gratifying note of menace in my voice.

The man nodded.

'Step away from the donkey and you can have it back.'

He shrugged and took a pace backwards.

'Further.' I heaved the child off its feet and it began to emit throttled sounds. 'Twenty paces. Count them aloud.'

He started to back away, counting monotonously. I lowered the child again, waited till twenty, then kicked its arse as hard as I could and it scuttled off into the darkness, muttering obscenities.

'If you want the other one,' I called after them, 'keep moving and keep counting – loudly. I want to hear you getting further away.'

'Twenty-three, twenty-four, twenty . . . one, two, three . . .'

The voices were dwindling. I walked back to the tree and nodded to Roland and Ellen to release the other child. Freed from their grasp, it twisted like an eel to spit full in Roland's face, then ducked for its knife and trotted briskly towards the fire which it demolished with a single swing of its foot, scattering embers and extinguishing the recently ignited branches.

As our eyes acclimatised again to the darkness, there was a soft grunt from Nebuchadnezzar, followed by a short gurgling cough. All went silent for a moment, then a peal of malicious laughter rang out and footsteps pattered away across the leaves.

'Oh no,' said Ellen softly.

Roland stood quite still and said nothing. I wondered briefly whether he had grasped what had happened. Then, with a despairing moan, he fled towards the donkey.

A sense of elation had risen within me once the danger had passed, but now it was obliterated by rage. A small, scrawny neck filled my mind, ripe for wringing. I made to move off in pursuit but Ellen restrained me.

'It won't do any good, Creb,' she said, 'and you'll only get hurt. Stay here. Build up the fire.'

I watched her walk towards Roland and reluctantly admitted that she was right. The damage was done and I could not repair it. She was right, too, about being alone with him. There were moments when an elemental comfort was needed. I waited a while, until the fire was blazing, then went to join them.

The stroke had been so swift and deadly that Nebuchadnezzar had not even had time to roll over. He had simply crumpled to his knees and remained there, his head lolling forward into a dark puddle which seeped slowly around his muzzle, glistening viscously in the firelight. Roland sat on the ground with one arm draped loosely across the broad grey flanks and stared blankly ahead. Ellen squatted beside him in silence. The drizzle had let up and now there was just the crackle of burning branches.

'Come back to the fire,' I said, helping him to his feet. We led him to the tree, sat him down and put a blanket around his shoulders. He said nothing and continued to stare into the darkness beyond. I studied him, looking for the fever-spots on his cheeks, the flutter of his hands. But he seemed quite composed.

Ellen asked him gently if he was all right.

'Yes,' he replied, without shifting his gaze.

Ellen and I glanced at one another. She shrugged and put a finger to her lips, suggesting that he was best left alone. We moved to the fire and sat down in silence.

Some time later Ellen returned from wherever her thoughts had taken her and pointed at my thigh. I looked down at the short rent in my breeches, a couple of hand spans above the knee. The wound had bled a good deal, but the hot pain had gone swiftly and now it merely ached.

'It's nothing bad,' I replied, then remembered the bloodied tip of the second urchin's knife. 'Are you cut too?'

She nodded and rolled back the sleeve of her cloak. Two dark trails ran down her arm from a ragged puncture near the shoulder.

'Does it hurt?'

'A little,' She pulled a face. 'But I'll mend you first.'

She reached for the pot and scoured it with a handful of leaves, then put a little water on to heat while I, feeling suddenly foolish, removed my breeches. I sat down, thankful for the long shirt-tails I could pull around me for decency, and turned my injured leg to the light. The wound was short but quite deep and I guessed that the blade must have reached halfway to the bone. Ellen produced a handkerchief, dipped it in the pot and began to dab away the dried blood. She clucked softly as the crust dissolved, exposing the lips of the wound.

'We should bind this up, Creb. Else it won't close.'

I had lost my concentration the moment she laid her hand on my thigh. Her skin felt cool and dry against mine and a soft musk crept from her hair, her neck and shoulders and the warm, shadowed places within her garments. It felt as if the intangible barrier that quivered between us was ready to fall away – and that we could not permit in Roland's presence, now least of all.

She was tying a knot in the strip of rag which now bound the wound, when the sound made us turn together.

Roland had begun to rock backwards and forwards, murmuring to himself. His hands were clenched at his side and his eyelids trembled.

Ellen glanced at me questioningly and I nodded, reaching for my breeches. We left the fireside and crouched down before him but he appeared not to see us.

His voice was low, almost throaty. 'Crebanellen . . . crebanellen . . .' I looked at Ellen and she began to mouth the word slowly, breaking down the syllables and pointing, first at me, then at herself. Comprehension brought a sharp twist in my gut. Ellen had blanched. But Roland stared straight through us. Now he had begun to shake his head.

'Crebanellen . . . crebanellen . . . rolanalone . . .'

His face tightened suddenly, as if he had run up against himself. He began to pound his thigh with one fist, muttering inaudibly. Then he gave a violent shudder and the seizure took hold.

It was a good deal more powerful than the two previous fits I had witnessed. Holding his head proved quite beyond me as he writhed and jerked and arched his back. It was as if he had become host to some savagely destructive entity that would rend him apart in its efforts to get out – an almost demonic force of which Ellen was visibly afraid. She hovered at a distance as I did my best to prevent the flailing body from damaging itself, but when the jaws clamped shut and the Adam's apple began to twitch and I yelled at her to help, she came forward and knelt on his chest, stilling him enough for me to wedge his head between my knees and eventually succeed in forcing the handle of a wooden spoon between his teeth.

Release came slowly. Long after his jaw had slackened, the spasms maintained an intermittent assault, as if he were in the hands of some absent-minded marionette master. It seemed an interminable time before he finally gained repose, his breathing becoming regular once more and the sweat cooling on his face.

We wrapped him in our blankets and moved back to the fireside, too cold and too overwrought to sleep.

'Is it always like that?' Ellen asked, the shock catching at her voice.

'No. I've not seen it worse.'

'Poor Roland,' she said softly, shaking her head. 'I suppose it was . . . Nebuchadnezzar . . . that caused it?'

'Yes . . . yes, it was.' Would she mention the other thing? I held her gaze for a moment but she said nothing. So we did conspire . . .

'It's a shock that does it,' I said.

'Poor Roland,' she said again. 'We'll have to get him another donkey.'

'Yes.'

She looked at the fire for a while, then: 'He worships you, Creb. You know that, don't you? As if you're the only friend he's ever had . . .'

'He had no life in the village,' I replied. 'He . . . frightened people – ignorant people. The steward was too afraid to let him be on his own . . . before I came along.'

'How did that happen?'

'Well . . . there was a murder . . .' I stopped, my mind dragging me back insistently to the lagoon. And as I peered again into the cold, black water, it struck me forcibly that I could no longer narrate one set of events without the other. Indeed, far from seeming unrelated, I now saw them melded into a single, sweeping stroke of fate. To ignore it would be to deny the truth of my existence.

So I began again, sketching in the features and the wandering path that linked them as I drew my invisible map for her, willing her to look through the space between us, to see the landscape as a whole and to understand . . . The way she looked at me as I spoke of my brother, sympathetically but steadily, reminded me that we were united, all three of us, in the experience of sudden and violent bereavement.

'Were you not afraid of him?' she asked when I had finished.

'Yes. At first. But then I understood that he only spoke what he already knew. And he knew that . . . just by . . . looking.'

I regretted the explanation immediately, sensing that now I had surely trapped us both, that the link between us was still too fragile to withstand acknowledgment.

'Anyway,' I said, clumsily seeking a way out, 'he's not always right . . .'

Ellen glanced away and began to roll up her sleeve, then looked back with the trace of a smile.

'Help me clean this,' she said.

We heated water again and she held out her arm; pale, slender and shadowed with a fine down. I dabbed with the handkerchief and she winced as it grazed the mouth of the wound, but nodded to me to continue. This dark disfigurement of the flesh seemed to heighten her vulnerability. The sudden urge to protect was almost overwhelming and when her breast brushed my arm, I felt my composure at last slide away and paused helplessly with the handkerchief in mid-air. I lifted my head to find her already looking at me directly and we remained there for an age, our eyes locked, before she murmured my name and leant forward to put her lips to mine.

Instantly the tension was swept away in a long, deep melting, a sudden wondrous contraction of the universe into a dark womb of touch, smell, sound and pure emotion. 'I love you, Ellen.' The words rose like clear water from a wellspring I did not know I possessed. I reached for her breast, soft and full beneath the bodice, and she murmured again, nuzzling into my neck. I allowed my fingers to trace downwards, feeling the fabric shift over the warm, firm flesh beneath, down across her stomach, down to her belly, the sudden change of texture . . .

'No, Creb.' She pulled away, shaking her head.

I cupped her face in my hands and pulled her gently towards me again. She did not resist, but her eyes were tight with anguish.

'What is it?'

She looked at me intensely. 'I . . . I can't . . . not now . . .'

I glanced towards Roland. 'Is it him?'

She replied with a curious little movement of her head. Whether it was a shake or a nod I could not be sure.

After a while she squeezed my hand, then rose to her feet and straightened her clothes. With a small, tense smile she said: 'We should sleep now.'

Eleven

acking the wherewithal to bury him, we covered Nebuchadnezzar with leaves the next morning while Roland still slept. I was ready to admit that I had never really liked him, nor properly understood the nature of Roland's affinity with him, but he had served a valuable purpose nonetheless and nothing, not even a donkey, deserved such a casual death. Anyway, I had grown used to him, to the routine he had imposed upon us with his plodding pace and regular pauses for feeding and watering. I found myself glancing with mild regret at the incongruously lumpy mound we had created on the forest floor.

Ellen and I returned to the fire and sorted the few possessions we had acquired – the cooking utensils, a couple of water-skins, some food – into a bundle that we could carry until we could find Roland another mount.

Roland continued to sleep and I was thankful for the persisting mild weather. Although I was still largely ignorant of his condition, I sensed that after a fit, when his whole system seemed to decline to an almost reptilian torpor, he was dreadfully vulnerable to the cold. So far we had been lucky with the nights we had spent in the open.

We sat around the fire, waiting for him to wake and talking inconsequentially. Some mention of Nebuchadnezzar called to mind the children of Judah and the fiery furnace. I asked Ellen if she knew the story, and she promptly unlocked a treasury of tales about the old kings and prophets with their outlandish deeds and still more outlandish names. Her familiarity with them was evident from the humour, the affection, and irreverence with which she told the stories, throwing her hands in the air, rolling her eyes, changing her voice to suit the different characters. It would have been a captivating performance had I not sensed all the while that she was really trying to disguise the tension which had returned between us. It was as if we had retreated from one another, and frustrating as it was, I lacked the confidence to do anything about it. So I sat beside her in suspense and, when she ran out of stories, turned the conversation to our journey so that I could seek what reassurance there might be in her look, her gestures, and particularly in her inflection of the word 'we'. Was there that subtle warming there, that refinement which excluded Roland, or did it still mean all three of us? I could not tell.

Midway through a sentence she suddenly stopped and shook her head from side to side with a deep frown. She reached across and clasped my hand and said fiercely: 'Oh Creb . . . last night . . . I'm sorry . . . it's – it's difficult . . . too difficult . . . I can't explain . . . please be patient with me . . . if you can.' She shook her head again, then looked up at me, eyes brimming with apology and frustration and confusion.

For a moment I said nothing. Then, despite myself, I asked: 'Do you want to stay with us – on the journey?'

'Oh yes, yes.' She nodded vigorously. 'I want to stay with you. This is a good journey for me. Good because I've chosen it, good because of you, good because of Roland . . .'

'I'm glad,' I said.

She smiled now and stroked my hand. 'Be patient, then?'

I leant forward and kissed her lightly. She did not pull away.

'Yes,' I replied.

An hour after setting off, we emerged from the woods and began to make our way through a tract of undulating farmland. There would be more forest soon; it was inevitable. I recalled hearing once that north of the broad valley where the town lay, a squirrel could cross the country from coast to coast without ever touching the ground, but for tonight at least we would be in relative civilisation, with the chance of finding decent shelter.

The sky had been dull all morning and now the drizzle returned, fine but persistent. Within a short time the damp and chill began to penetrate, and at Ellen's suggestion Roland and I took a blanket each for protection, while she pulled up the hood of her cloak. There was a certain severity in the way it framed her face, accentuating the soft curve of her cheeks and the large brown eyes. It made her, I thought, more attractive than ever.

I glanced at Roland, who was on the other side of her. He was pale and subdued, as I had known he would be, but he seemed to be coping well enough with the slow pace we had set, and he had made no mention of Nebuchadnezzar, pausing only to glance at the mound of leaves as we left our camp. If he was still distressed he was concealing it well; as for the other matter, I prayed that amnesia had taken its customary effect.

The land around us now was open and well tended, a comforting ripple of low grassy hills and ploughed earth, neatly turned in furrows of rich, dark loam. Here and there wedges of green indicated a winter crop. A cottage or two stood sheltered by a coppice, while a ribbon of stream led to a distant water-wheel, turning slowly beside its mill. After the confinement of the forest we should have felt cheered, yet the grey light, the drizzle and the bare trees made the landscape seem almost desolate, as if winter had sucked the life from it.

For an hour or more we saw no one, and again I found myself wondering about pestilence. The condition of the land suggested that it might have passed by, but the absence of livestock, or indeed of any people on what was clearly a much-travelled road, pointed to the contrary.

In the early afternoon the drizzle eased and the sky began to clear. For a moment, watery sunlight splashed a hill, half a mile ahead of us, and we could see the scar of the road, curling across its flank. A short while later we paused at the crest and found ourselves looking down on a village, beyond which extended a long shallow valley. Bushes edged a stream running straight down the centre of the valley to the point, a mile or so beyond the village, where it swept out on a detour around the foot of a lone hummock.

The hummock rose distinctively, a curious disfigurement of the smooth valley floor, a kind of goitre almost, looming a hundred feet or so above its surroundings.

Ellen, who had been trailing a little, now came up behind us, paused and looked for a moment, then flung an arm around each of our shoulders.

'Well, well, well!' she exclaimed.

I had already made the calculation. Allowing for the slower pace today, now that we were donkey-less, it would be about half as far again from the hanging-tree as the tree was from the castle.

I nodded.

'It does look . . .' Roland began, but his words disintegrated into a fit of coughing. Ellen fumbled for a water-skin. He drank and the irritation subsided, leaving him red-faced and watery-eyed. He cleared his throat. 'It does look extraordinarily like the one on the map.'

If the resemblance was uncanny, so, in a sense, was the hill itself. A fairy hill, perhaps. I had never seen one but I had heard people talk of them. Grassy, perfectly rounded, built by the little folk for their convocations, with a soft dome of turf on which they could dance of an evening. A place to be wondered at by day and avoided by night.

We rested a while, drinking in our apparent good fortune, then began to make our way downwards towards the village, skirting the large, well appointed manor which stood, surrounded by beech and elm, at the foot of the hill.

The village was a pleasant looking place. The cottages were clean and neatly thatched, there was little refuse to be seen and the pond in the centre, fed by a fast flowing brook, was deep and clear. Downstream, the brook had been diverted into a second channel for watering animals. An orderly hand, it seemed, had reached out from the manor.

The sun emerged again, gilding the single willow overhanging the pond, and the place seemed to stir with a presentiment of spring. But still there was no one about, nor any trace of animals. There was a strange feeling here, a sense not so much of abandonment as of containment – a damped fire, rather than one that had been left to go out. The village was still inhabited; I was certain of it.

We walked to the pond and paused, looking around us. All was silent. Doors and shutters were closed, but faint smells lingered – woodsmoke, cooking, curing leather, ordure and something rich and fatty I could not identify. Ellen gave an exaggerated shiver and crossed her arms, but Roland shook his head.

'I believe they're *asleep*,' he said incredulously.

I nodded. Slowly, the same thing had been occurring to me.

'But it's only late afternoon,' Ellen protested.

Roland began to say something but a spluttering cough intervened. We stood still for a moment. Nothing stirred.

'Could they be sick?' whispered Ellen with a grimace. 'Could there be pestilence here?'

Roland breathed deeply then shook his head again.

'Then what is it?' I asked. 'Even if they are sick, where are the animals?'

'We could,' said Roland slowly, 'wake someone up and ask them.' There was a glint in his eye.

Ellen shivered again, but now Roland was intent on our surroundings. After a while he flung a jubilant arm towards the opposite end of the village, where a tendril of smoke drifted up from an unseen chimney. He set off immediately, leaving us to follow.

At the very edge of the village stood a cluster of cottages, from one of which the smoke emerged. The track wound past them, dropping gently into the valley beyond. Fifty yards on from the last cottage stood a larger building, half-timbered and with an upper storey. A broomstick extended rakishly above the front door, signifying that the place was an alehouse with its suggestion that cobwebs could be swept away by the good company and beverages to be found within.

Still a little way ahead of us, Roland hesitated for a moment, glancing between the cottages and the alehouse, then made up his mind and strode purposefully down the track. By the time we caught up with him he was standing with his fist raised to the door and an expectant look on his face. He knocked again and after a long while we heard a faint shuffling of feet, then a voice: 'Who's there?'

'Travellers,' said Roland loudly. 'We need food and lodging – if you have it.'

There was a long pause, then: 'Still light, is it?' The voice was curiously muffled.

Roland glanced at me, raising his eyebrows. 'It's still light.'

'Food and lodging, you say?'

'Yes,' Roland replied. We waited. Roland raised his voice: 'If it's too much trouble, perhaps you would oblige by telling me how far it is to the next place.'

'Half a day,' came the eventual reply.

We looked at one another and I shook my head. 'That's a little far,' said Roland. 'We'd be pleased if you could take us in – unless there's pestilence . . .'

The response was emphatic. 'No pestilence here, my friend.' A ruminative silence followed. 'Very well. Come back when it's dark.'

The feet shuffled off again, leaving us to exchange incomprehending glances. Ellen now looked nervous. She began to protest but Roland intervened, nodding to himself: 'No harm'll come to us here.'

'But how do you know?' she persisted.

He merely shrugged and spread his hands. Ellen looked at me, suddenly vulnerable again, and I longed to put my arms around her. I reminded her that Roland was usually right in these matters, and that we had little choice in any case. She nodded, holding my eyes for an instant.

For the remaining hour or so of daylight we occupied ourselves by walking out to the fairy hill. It was so perfectly domed and so evenly covered with short, springy turf that I was sure we must find a tiny doorway set into its side. But there was none. It was merely as the map had suggested – a grassy hump of earth, significant for nothing more than its shape.

The map, the damned map. So accurate at one moment, so unclear at the next. And so casual in its treatment of the landscape. No sense of dimension or relief, texture or shape; just an impossibly straight road running through an impossibly flat and featureless terrain. The product, I concluded, not only of a secretive mind, but of an unobservant eye and a thoroughly undraughtsmanlike hand. Yet we were permitting it to determine the pattern of our existence for the foreseeable future. Or was that to credit ourselves with a choice in the matter which we did not really possess? Was there, in that sheet of parchment with its tracings of ink, some talismanic power? And if there was, what difference would it make? For what else might we be doing, adrift and unskilled in a troubled world?

As dusk settled and we made our way back towards the village I felt increasingly fogged. It was not only the map that troubled me, but where I stood with Ellen, with Roland and ultimately with myself. Nothing seemed clear. I was badly out of sorts by the time we reached the alehouse again and halted under a clear, cold sky, speckled with stars.

It was tallow, that indefinably rich and fatty smell I had caught as we had first reached the village. Stepping inside the dingy, raftered room of the alehouse was like entering a tallow-works.

The air was greasy with it. Pools of it lay congealed on table tops. Tongues of cold tallow slid down the necks of bottles. It spat on the fire as it dripped, melting from the chimney-breast. And the host, a nondescript middle-aged man in a leather apron, had a complexion the same colour, waxy and pallid, almost corpse-like.

His wife, tall and angular, was moving around the room with a taper, attending to the candles which forested every available surface. They guttered and flickered as they caught; the whole place swayed crazily with shadows.

The host fiddled with a wisp of grey hair as he told us that he could feed and lodge us, but that we would have to take what came and that he only had one small room. He paused, assessing us, then added cryptically: 'Can't promise you'll sleep. An' I want you gone 'fore sun-up. Payment in advance.' He named a small sum and held out his hand. I paid him and he directed us to the fire.

'There'll be food in a while,' he said, disappearing after his wife into the nether parts of the house.

We took our seats. The candles burnt steadily now and the place was ablaze with light.

'They must be expecting a lot of trade,' said Ellen with a little frown.

Roland shook his head thoughtfully. 'Look at the windows.' Across the apertures, concealing the closed shutters, hides had been stretched taut and batoned to the woodwork. Propped against the wall by the doorway was a hide-covered, door-sized frame. Pegs had been driven into the jambs. Corresponding holes had been drilled in the under-surface of the frame.

'This room has seen no daylight for months.' Roland sniffed. 'Nor much fresh air.'

The newly-lit candles had burnt off a little of the tallow-smell. It would return again as they began to melt, but for the time being the air bore a flatulent seasoning of humanity and sour ale.

Roland tapped his stool with his fingers, still pensive. Eventually he looked up. 'I believe they're nocturnal . . . like owls. Asleep in the day, awake at night.'

Ellen pulled a face of mystified amusement and spread her hands. 'But why?'

Her question was curtailed by the entry of the woman with food. As she approached the table I studied her more closely. Her movements were stiff, a little jerky, as if some essential lubricant was lacking; the pallor of her skin hinted at the sluggish passage of blood through her veins. The life-force did not seem abundantly present here.

She placed the tray heavily on the table and moved away without a word.

'Might I ask a que . . .' Roland coughed again, reached for a cup and drank deeply.

The woman turned, her eyes narrowing. 'Is 'e ill?'

'No, no.' Roland dabbed at his eyes and drank again. 'Just a tickle.'

'Ask away then,' said the woman. Her expression offered little encouragement.

Roland cleared his throat. 'Do you . . . er . . . sleep during the day?'

'What's it to you?'

'Nothing, nothing . . . merely curious.'

She hesitated a moment, then nodded curtly.

'Everyone here?' He gestured to encompass the whole village.

Another nod.

'On account of the pestilence?'

Caught off guard, she stared at him for a moment, a faint twitch at the corner of her mouth. Then she recovered herself and replied stolidly: 'Our lord watches 'eavens. Says there are spots on the sun. Plague-spots. An' now pestilence's come down to earth on the sunlight, warm air, some such thing. So we close up the doors and windows. Only go out at night, we do. Live our lives at night.'

'And . . . no pestilence?'

She shook her head.

'But what of the animals? Can they be made to . . . change their habits?'

She shook her head. 'Slaughtered 'em. Kept some chickens, o'course, a few pigs, couple of milk cows. It's winter. They was inside anyway. Slaughtered the rest. There they are.' She waved at the candles, the hide screens, our plates of food.

'And the land, the crops, the vegetables . . .?'

'Makes no difference to 'em. Work at night, by torchlight.' She was beginning to appear irritated, but Roland held her eye and continued: 'And how long will you go on like this?'

'Till 'e says it's safe to stop.' She shrugged. 'We're used to it now. An' spots're goin', anyway. 'E watches 'em every day wi' some kind o' glass. Keeps a cloth o'er 'is 'ead.' She shuffled her feet, began to edge away.

'Are you not afraid to take in travellers?'

'Why should we be? Don't get pestilence from no people. Only sunlight.' She turned peremptorily and left the room.

'Extraordinary,' said Roland, shaking his head. 'I don't think she believes it, either.'

'Why go along with it, then?' I asked.

Roland looked blank.

'Because everyone else has,' said Ellen. 'I wouldn't trust her.'

Roland shrugged, as if the woman's trustworthiness was of little consequence, then shivered and pulled his stool closer to the fire. We set about our food. In due course we began to hear muted sounds of activity outside – voices, the measured hoofbeats of a large horse, the rattle of carts. From time to time a pungent smell of burning resin sharpened the tallow-laden air as the door creaked open and sleepy looking customers drifted in to consume their food with a silent concentration which suggested that the day's affairs still lay ahead of them. Those who spared us a second glance did so with a hint of envy, I thought.

I stretched, luxuriating in the warmth of the fire and the feeling of hot food settling in my stomach. Ellen and Roland both looked flushed and seemed to have drifted away with their thoughts. I found my own turning to the last time we had been in an inn; to Ellen and her uncle, to the look, like that of a trapped animal, which had dulled her eyes as she left the table. Despite myself I began to imagine what had taken place in some out of the way corner of the town: Ellen stoically containing her disgust as a man, to whom I could put no face, panted and fumbled in the darkness, his hands free to explore the body that mine had been denied . . .

As the image lingered I was stung to anger, protectiveness, a sense of possession. But in some perverse and unbecoming way, it also aroused me, and with arousal, frustration returned.

Roland rose from his stool, announcing that he was going to bed. I asked how he felt and he frowned briefly, then gave a vague smile. 'Tired.' He left the room and we heard his footsteps on the stairs.

For a while we sat by the fire in silence. Ellen seemed almost to have forgotten my presence, but at length she turned to me with a curious look, both wistful and forthright, and I felt her hand slide across into mine, firm and warm.

'Creb . . . this may sound shameful,' she began, 'but . . . I've . . . always liked men. I'm . . . good with them, you see. I knew it when I was quite young . . . too young, I expect.' She smiled. 'But it was innocent pleasure . . . and they weren't really men anyway, just boys. It was almost a game. They enjoyed it. So did I. And then a young man came along. A nice young man. A Jew, like us. He was my father's apprentice. I saw him every day. He brought me flowers, kissed me sometimes – when my father wasn't looking. Oh, and I egged him on, no doubt. He said he loved me. And the strange thing was that I didn't want him in the least, yet I thought I loved him, too.' She paused and looked down. 'When they . . . burnt him . . . along with all the others, I thought my heart was broken – literally broken, as if there were pieces of glass inside me. It hurt so much. I knew I'd never be able to love anyone else.'

She must have felt me wince, for she squeezed my hand again and continued with a little smile: 'But you know, Creb, it didn't last very long – the hurt, I mean. I forgot about him quite quickly. In fact I'd scarcely thought of him for weeks until . . . recently. You see, it wasn't really love – any more than it was with the others. It was just that with him I was. . . . flattered, unused to being paid so much attention.' She paused. 'But there was my uncle and . . . all of that . . . that helped me forget about him, drove him out of my mind. Drove everything out of my mind, I suppose.'

This was so different from the way she had first told her story. Then she had seemed almost unnaturally detached. Now she was animated, connected again to herself, to the living memories with which she conjured.

As she went on, the rancour flared in her voice: 'My uncle was a swine, Creb – I hope you hurt him, back there. He deserved it. He knew I was no virgin, not like some of the other girls. And he used it – oh, don't ask me why – maybe because he was too idle to work himself if he could get someone else to – maybe because he didn't like women. But he used it – and . . .' she shook her head 'and . . . soiled a part of me . . . in the doing. It was almost like witchcraft – turning something that had always given me pleasure into something vile and detestable.' She gave a wry laugh.

'Anyway, I've learnt much since then – hope, for example – or maybe you would call it survival . . . how even when things are as bad as they possibly could be there's something inside that makes you keep believing they'll get better. And people. How you come to tell so quickly whether they're good or rotten . . .'

'You sound like Roland,' I said.

'No.' She looked at me intensely for a moment. 'Like you, Creb. Roland does it naturally. He always has done. He doesn't even have to think. But you've learnt, like me – the hard way.' She paused reflectively. 'Roland's still . . . quite innocent. He knows all sorts of things – by instinct. But he doesn't really know how to put them together, how to *use* what he knows.' She smiled again. 'Oh, don't misunderstand me, Creb. I like Roland as much as I've ever liked anyone. He's suffered too. And he's not just clever. He's generous and brave – very brave, I think, in his own way. I've never met anyone like him before. But he hasn't really grown up yet.'

The candles were burning down and in the gentler light Ellen's eyes seemed huge and vibrant. I wanted to run my finger over her cheek with its soft bloom.

'But you've made sense of things for yourself,' she continued. 'I knew it the first time I saw you. And I knew something else then, too.'

'What?' It came out in a half-whisper.

'That I could love someone else. That the feeling I thought I had lost forever was not the real feeling at all. That it had been just – a girlish thing.' She paused. 'Whereas this was something quite, quite different.'

I leant forward and clasped her hands in the warmth of her lap.

'So did I, Ellen,' I said. 'I mean . . . knew it was something different . . . with you.'

She looked at me with extraordinary tenderness. 'How can such a wonderful thing happen, Creb?'

I hesitated, then shook my head. 'Best not even to ask.'

For a long time we looked at one another, close together, oblivious to our surroundings. Then she lowered her eyes and began to stroke my hand as she continued: 'I told you all that – just now – so that perhaps you would see why last night was – difficult. Maybe not understand, but at least see. It has nothing to do with whether I love you or not. I suppose it's just that I'm – well – a part of me's still . . . under his – my uncle's – spell, and . . .'

'You don't have to say more,' I said. 'I understand.' The urge to gather her up and carry her to somewhere soft and warm, to undress her, to kiss and caress and smell and breathe her, was almost overwhelming.

But she continued as if she had not heard me.

'And now . . . last night, I mean . . . it was the first time since . . .' she had begun to tremble 'and I wanted you so much . . . but as soon as you touched me . . . in my head . . . it was as if . . . as if I was drowning in a black, cold sea.' Her voice was lost in a sob.

'Oh, Ellen.' I tilted her face toward me. 'I can wait.' I said it as gently as I could, but I knew I had failed to conceal the edge of frustration. She was too caught up in her own anguish, though, to notice.

'But I don't want to make you wait, Creb. I don't want to wait. I don't want anything in the way of us.' She buried her head in my shoulder, then looked up again and whispered fiercely: 'Let's go upstairs. Now. Roland'll be asleep.'

We rose and left the room, clinging to one another like the survivors of a shipwreck.

Twelve

oland was not asleep. As we entered I heard him turn and give a shivering sigh.

'Did you bring the blankets with you? I'm very cold.'

The room was tiny. It was also draughty and dark, but not completely so. The reason for this was quickly apparent. It had not been sealed with hide screens like the room downstairs, and although the shutters were closed, they were warped and ill-fitting. Since it was used, presumably, only at night and then only by passers-by, the host had not bothered with his precautions here.

Four straw mattresses lay side by side with a few inches to spare around them. Roland was on the one furthest from the door. They crackled as I stepped over to him and I realised gloomily that Ellen and I would never have been able to make love without waking him. The only small consolation was that he had left us no choice but to lie next to one another.

I knelt down and laid two of our blankets over him, noticing as I did so that he was already wrapped in the threadbare coverings from the other three mattresses as well as his own.

'Thank you,' he said, burrowing down beneath them. 'I couldn't sleep. Too much coming and going outside – people wandering about with torches – carrying on as if it was broad daylight.' He tried to laugh but was shaken by a dry, shallow cough.

Ellen asked if he was ill. He gave a little shrug and replied that a drink would help. I fumbled in the gloom for a water-skin, then turned back to see that Ellen had her hand on his forehead. She glanced at me, mouth downturned.

Roland sipped, spluttered and coughed again. 'It's . . . uhuh . . . just . . . uhuh . . . a touch of fever . . .'

'I expect so.' She removed her hand. 'Are you shivering?'

'A little.'

'Anything else?'

There was a long pause.

'Tell me what it is, Roland,' she said gently.

'Well . . . I have a pain. Here.' He flapped the blankets in the middle of the bed. 'In my groin. More of an ache, really. It's tender.'

'Put your hand there. Can you feel anything? Is it swollen?' She was calm, encouraging him like a child.

He fumbled beneath the blankets. 'A little, maybe. It's hard to tell.'

'From all that walking we did today, I expect. Now, we need to keep you warm.'

She tutted to herself and turned to me. 'Creb, I must have left my cloak downstairs. Would you fetch it?'

I stepped over the cloak, lying where she had dropped it in the gloom by the door, and left the room, went noisily downstairs, then tiptoed halfway up again and waited. A minute later she joined me.

'He's very hot, Creb. He certainly has a fever.' She paused. 'But that's not all. The ache . . .' She caught my wrist and looked directly at me. 'Do you know what a bubo is?'

The word was vaguely familiar. I shook my head.

'A bubo,' she said deliberately, 'is a boil in the groin or under the arm. It's one of the symptoms of pestilence, Creb.'

I searched her face but the expression was unequivocal.

'Mother of God!'

She nodded, letting go of my wrist. 'Now it may be that he's not used to the walking – or that he twisted himself during the fit. And the fever may be just that – a fever. Let's pray it is. But he's sick, however we look at it, and I doubt he'll be well again by tomorrow. I think we must take him away from here.'

'Now?'

'Now.'

'Could we not wait – at least till morning? If it's just a fever, then isn't this the best place he could be? It's warm, there's food . . .'

She shook her head vehemently. 'No, Creb. That woman. Roland was right. She doesn't believe all that claptrap about sunlight. And the way she looked at him when he started coughing . . . If she hears him now and thinks he's ill – with anything – she'll raise the roof. And it wouldn't be good to have him upset as well as ill, not after last night.'

'What do we tell him?'

She looked at me. 'You can't tell Roland anything but the truth. You know that.'

I nodded sheepishly.

'But not all of it, Creb. Not yet, anyway.'

I hesitated for a moment. 'Ellen . . . what if it is . . . the other thing?'

She shook her head. 'I don't know . . . let's wait and see.'

With a brief smile she took my hand, and led me back up the stairs.

We entered the room to find Roland sitting up, his pale face now peering absurdly from within the hood of the cloak. Below it, his body ballooned with blankets like an overstuffed scarecrow. He was shivering visibly and I thought I could see traces of sweat on his face, but he attempted a grin all the same.

'You told me to keep warm.' He looked from Ellen to me and back again, still grinning. 'So? I know – it's pestilence . . .'

'Roland!' Ellen scowled convincingly. 'You are feverish, though, and you're likely to cough all night – and to be truthful, Creb and I are worried about the woman downstairs . . .'

'What – that she'll hear me, make a to-do, throw us out?'

Ellen nodded.

'So we should go now?' The grin slipped. 'Yes. Perhaps we should.' He started to rise, became entangled in the blankets and fell over, coughing like a sick sheep. I passed him more water.

'Anywhere particular in mind?' He began to search for his boots.

Ellen and I looked at one another.

'Oh well . . .' He stood up unsteadily. 'Perhaps a walk will help me shake it off, whatever it is.'

The moon was out and the air sharp as we dropped down into the little valley beyond the village. Ahead of us, the fairy hill rose from the shadows, smooth and round and faintly silvered with frost.

I began to imagine myself standing on the summit in the moonlight. Perhaps the fairies were calling to me, perhaps I just needed the balm of higher ground for a moment . . . I walked briskly ahead of Roland and Ellen and climbed the frosted mound, feeling not the least afraid.

I stopped at the top and turned to watch my companions approach along the shadowy valley floor – small dark figures from whom I was momentarily detached, yet bound by the invisible ties of friendship – no, more than friendship, love. I squatted down, feeling the chill rising from the ground beneath me. Earth. The map was right there, at least. This was earth in its most perfect representation. Solid and shapely, unmoving, age-old, reliable. The foundation upon which all things could be built. Constancy, even when the world around had gone mad. And madness, or something akin to it, was close by – the glowing points of torchlight, their bearers unseen in the darkness, moving hither and thither through the fields around the village. It reminded me of the scene at the manor when Roland and I had paused at the edge of the forest, I full of anger and he of bewilderment. Love, constancy, responsibility . . . Something still remained to complete the circle of my thoughts. Patience, that was it. Oh, Ellen. A brighter star than any that winked through the frosty darkness. I would be as patient as the hill beneath my feet.

I walked slowly down again as Roland and Ellen drew level and took my place between them, an arm around each. Roland shivered continuously beneath his swaddling of blankets, his face drawn and ghostly in the moonlight. I asked him if he could keep going and he tried to answer, but coughed instead and nodded briefly. Ellen's arm stole around my waist.

We continued along the valley, scanning the darkness for shelter, but there was nothing more than the occasional clump of bushes, and no habitation of any sort. The stream chuckled beside the path, a fine film of ice beginning to glint at its edges. Our movement kept us warm enough, but Roland was leaning more and more heavily on me. Ellen went to his other side so that we could support him between us.

A little further on, the valley tapered into a gulley between high grassy banks. Something was standing there, beside the path. We drew closer to find a dozen shoulder high staves, festooned with ribbons of rag and planted in the earth beneath a niche dug into the side of the bank. The niche had been lined with slabs of stone. In its centre sat a crude wooden carving of a man striding forward with a staff in one hand and a tiny child perched in the crook of his other arm. It had once been painted, but had long since weathered, leaving only incongruous slivers of pink on the child's fingers and a speck of blue in one of the man's eyes, echoed in places on the waves lapping at his knees.

It was a drab little scene, the statue cold and lacklustre in its stone alcove, the rags hanging limp and rotten from the staves. But we stopped all the same.

'Your namesake,' Roland wheezed.

Ellen, who had produced her crumpled, bloodied handkerchief and was now tearing off a strip from its edge, paused and looked at him quizzically.

'Creb – Christopher. Did he not tell you?'

She smiled and shook her head, then leant forward to tie her offering to the nearest stave.

'He's . . . part of your faith?' I had noticed how instinctively she had reached into her cloak.

'No. But we should always seek help where it's offered. It doesn't matter how.'

Roland muttered something in heartfelt agreement and fumbled inside the blankets. I took out the knife and helped him to remove a piece of shirt-tail, then did the same with my own.

We added our offerings to the throng and paused for a solemn moment before moving on.

An hour later we came upon the monastery. Clouds had drifted in now, and we were aware only of a shadowy bulk ahead of us, squatting in an unkempt meadow. We could hear the river beyond, into which flowed the stream we had followed all the way from the village.

Ellen muttered some word of thanksgiving in her own tongue. Roland grunted and I felt his weight lessen a little on my arm.

As we drew nearer and the buildings began to detach themselves from the shadows we were confronted by what at first appeared to be a small manor. Closer scrutiny, however, revealed a sharply pitched roof, crowned with a wooden belfry, and a simple cloister running along one wall. The cloister gave onto a cobbled court-yard, contained on the other two sides by long, low buildings. A little apart, but facing the courtyard, stood a small dwelling, no more than a cottage.

We were almost upon it when the moon broke through and the shabbiness of the place was revealed. Weeds sprouted amongst the cobbles, a hole gaped in the roof of one of the side-buildings and next to the cottage with its blackened, lumpy thatch, a vegetable patch had run to seed. The occupants had evidently departed. Not even a lingering sense of godliness remained.

We made for the main building, the church, but as we reached the arched doorway Ellen stopped and shook her head with sudden vehemence. We turned and walked into the courtyard where Roland was overtaken by a spasm of coughing. We paused to let him recover and as the echoes died, a door opened in the building to our right and into the moonlight shuffled the figure of a monk, stooped with age.

He made his way towards us uncertainly and halted a few paces off.

'Good evening,' he said without looking up. 'What business have you here?' The voice was cracked. Strands of white hair fringed his tonsure and swollen-jointed fingers trembled at his waist. But there was a tangible authority in his presence, some inner spark which defied the drabness of its surroundings.

'We need shelter,' I said, then added impulsively, 'our friend is sick.'

'Yes, yes,' said the old man. 'I heard.' He turned and shuffled back towards the open door. 'Follow me.'

It was pitch dark inside, but the monk seemed to have no need of light. Reaching out every so often to steady himself, he made his way down a corridor from which we sensed, rather than saw, openings to a number of rooms. At the end the corridor bore left, passing a low, arched opening and a glimpse of what appeared to be a kitchen, lit by a shaft of moonlight streaming through the roof. The old man opened the door beyond, admitting us to a darkened rectangular room. A chink in the shutters of its single window allowed us to see furnishings of two trestles, twenty or so stools and, at the opposite end, an empty fireplace.

'A little draughty,' he said, without apology, 'but you will want a fire and the dormitory is without means of heat. You will have to find wood, though. I seldom have need of it myself.' He paused. 'Leave your friend with me. I will take care of him.'

We led Roland through the gloom and sat him down on a stool by the fireplace, unburdened ourselves of our belongings and then, taking a blanket each, went in search of wood. There were trees a little way along the river bank. No one had gathered the winter's fallings and we filled both blankets without difficulty.

As we lugged them back to the monastery, Ellen said: 'Your namesake did us well, didn't he, Christopher?'

I stopped, startled by the use of my proper name. No one had called me that for years, not since my brother had begun to talk.

She caught my surprise and smiled. 'Chris-to-pher.' She said it again, slowly, softly, as if she were caressing the word with her lips. It sounded like the rustle of silk. 'Do you like it?'

'When you say it, yes.'

She leant across and kissed me. 'Good, so do I. We'll save it for when we're alone.'

We walked on and a sudden thought intruded: there was one thing that would grant us our solitude . . . I was immediately horrified at my own ignobility, at the blackness of some corner of my heart, but I permitted the thought to linger until my conscience pricked me so sharply that I said: 'He might be able to help Roland – the old man. Monks have all kinds of remedies, don't they?'

'Mmm . . .'

'No?'

'It depends on the ailment. Fevers – yes. I believe they do. But pestilence . . .' She sighed. 'I just don't know. I never heard of anyone being cured in our town, by monks or priests or anyone else. Oh, there were quacks who made up salves and potions which might have made the pain a little easier to bear. I told you about the apothecary – he was one – but I can't think of a single person who didn't die eventually. I suppose if it is pestilence, the best he could do would be to pray. At least he knows how to do that.'

'And what about us – if it is pestilence?'

'Ask him to pray harder.'

'So you do get it from other people?'

'Oh yes. If you'd seen what I saw, you'd know. The poor people – in the crowded areas – they died like flies.'

I felt suddenly empty, afraid not so much of death itself as of what it would steal from us. I halted, letting go of my bundle, and took her hands.

'I'm not ready for that.'

She looked at me for a long time, her eyes luminous in the moonlight.

'Nor am I, Christopher,' she whispered, 'nor am I.'

The refectory was lit by candles when we returned – tall, straight church candles made of beeswax, which burned with a clean, steady flame. A mattress had been conjured up from somewhere and Roland was now stretched out, coughing and shivering beneath a pile of blankets. At his side knelt the old man. He lifted his head as we came in and I was instantly struck by his eyes. They were clouded, almost milky, and utterly without focus. I wondered how he had managed to light the candles.

'Do you have wood?' he asked.

'Yes,' I replied.

'Tinder?'

'Yes.'

'Make the fire then, and boil water.'

Even if Roland had been at death's door, this was one task I would not have hurried – I had fumbled the lighting of too many fires in my time and had long since learnt that precision was all. I peeled bark from the driest of the wood, laid it on the bed of cold ash, for the fireplace had not been cleaned since its last use, then built up a pyramid of twigs and applied the flint. At the second attempt the tinder caught and a flame curled up, licking at the twigs. The gentlest puff, a sudden flaring, and the fire began to crackle, sending a thin column of clean, woody-smelling smoke up the chimney. Half buried in the ash were a couple of charried log-ends which I placed on either side of the pyramid, then laid larger sticks across them, forming a lattice around the summit of the pyramid to draw the hungry little tongues of flame upward and outward.

Behind me, Ellen and the old man were talking quietly. Then I heard her rummage for the cooking pot and leave the room – to fetch water from the stream, I imagined. I sat back on my heels, temporarily absorbed and ready to savour the moment when the fire congratulated me on my craft by consuming what I had built, then collapsing to create a good strong heart. It was a time for good strong hearts.

When Ellen returned, there was a steady blaze in the centre of the hearth and, on one side, a bed of embers between two stones.

She knelt down beside me, placing the pot on the stones.

'He doesn't think it's pestilence.'

'You told him?' I laid sticks on the embers and blew.

She nodded.

'Merciful God! How does he know?'

'He nursed them – the other monks – all of them. He's the only one left.'

'Does he know what it is, then?'

'Something in the chest. It could be lung-fever.' She shook her head. 'That's bad enough.'

I knew. It had killed my mother's sister.

'Can he do anything?'

She spread her hands. 'He says he has a remedy. That's what the water's for . . .' She turned towards him and raised her voice. 'It's nearly boiling.'

The old man heaved himself up, took several paces along a line parallel to the fireplace, then turned at right angles and made his way slowly but unerringly for the door. Feeling suddenly discourteous, I went after him and caught his arm.

'Shall I come with you . . . I mean . . . can I help?'

He shook his head. 'Thank you. I know every inch of this place. And you would only come to harm in the dark.' He paused. 'I should be grateful, though, if you did not move the furniture, or tell me if you have. The novices used to put stools in my way.' A wry smile crept through the wrinkles. 'Things are . . . simpler . . . now that I am alone.'

A short time later he returned carrying a small leather pouch, a stone pestle and mortar and a pewter cup. He sat down by the fire and, with movements of a quite unexpected deftness and precision, took three pinches of dried herbs from the pouch, placed them in the mortar and began to grind. As he gripped the pestle, the dry, parchment-like skin across the backs of his hands tautened until it seemed almost translucent.

'How long have you been alone?' Ellen asked as the herbs flaked, then began to fragment into a coarse powder.

He paused and raised sightless eyes to the ceiling. 'Three, perhaps four months,' he said.

'And you live here without heat, with no one to help you?'

He nodded. 'It is more than fifty years since I took my vows of poverty. I ceased feeling the cold after a year or two – oh, and I can still light a fire if needs must. Nothing so very much has changed. Except that I can no longer see.'

'But what about food . . .' I asked.

Roland was shaken by a paroxysm of coughing. Ellen moved over and laid her hand on his shoulder.

'Food . . .' said the old man, when we could hear ourselves once more. He smiled again. 'Self-sufficiency has its limits, I admit. There is a village a little way away. They are kindly people. They bring me provisions from time to time.'

He removed a pinch of powder and rubbed it between thumb and forefinger, assessing its consistency. Satisfied, he tipped the mortar's contents into the cup and motioned for water. As Ellen poured, the water turned amber in colour. Grains of undissolved herb floated to the surface and a sharp, almost metallic smell, not unlike that of cow-parsley, rose with the steam.

'Must you stay here?' she asked.

He stirred the cup with the pestle. 'I do not have to. But I choose to – or rather, Our Lord chooses for me to. I believe that is why He spared me. In any case, I know no other place, and there is much still to pray for.'

He tested the potion with his finger, then set the cup down and heaved himself upright.

'Let it cool a little, then see that he drinks it all. It will ease the cough and help him sleep. There is something else I may give him – for the lungs – but it is very strong, and we should wait to see if he needs it. I will be in the dormitory.' He set off on his course for the door.

'Won't you stay and share some food with us?' I called after him. The long walk, albeit in the middle of the night, had made my stomach rumble.

He shook his head. 'It is late and I have eaten today. Keep him warm.'

Ellen helped Roland sit up and I passed him the cup. He was flushed with fever and looked a little dazed, but he managed a grimace nonetheless as the steam reached his nostrils.

'It smells vile,' he said, then the coughing shook him and he had to pass me back the cup for fear of spilling it. At length he was able to take it again. He drank the potion in one draft, spluttered disgustedly and slid back beneath the blankets.

'Would you like something to eat?' I asked him.

He shook his head. 'Sleep – if I can.'

'You will,' said Ellen. 'The drink will help you.'

I built up the fire while Ellen diced a turnip and placed it together with half an onion, the last of our provisions, in the pot. We would have a little now, she said, and leave the rest for Roland in the morning.

As we waited for our meagre soup, Roland continued to fidget and cough and sigh like a restless animal, but eventually the potion, whatever it was, began to have its effect and by the time we had finished supping he had slipped into sleep, his breathing laboured but regular.

'So although everyone who catches pestilence dies from it, not everyone catches it,' I said, thinking of the old man. Ellen had stretched out with her head in my lap, watching the fire.

'It seems so.' There was a pleasant dreaminess in her voice.

'I wonder why.'

She shook her head and her hair rippled across my thigh. 'Who knows? Perhaps because it's ordained, like the old man said.'

'He must be lonely.'

She thought for a moment, then said: 'It seems a lonely faith to me.'

'Lonelier than yours?'

'Oh yes. For us it was . . . as much a part of our lives as going to market. And just as ordinary in some ways. It was really part of everything we did – our eating and drinking, our laughing and singing, the way we greeted one another. Oh, we couldn't worship or celebrate as openly as we would have liked . . . but despite that – or perhaps because of it – we felt a very special sense of belonging. It gave our lives . . . something else, I think.' She had become animated as she spoke, but rather than move her head from my lap she allowed her hands to convey her feelings. The slim fingers darted and danced in the firelight.

'But you Gentiles,' she went on, 'it's as if you keep it all in a little box and only let it out in church – or on your knees, alone at your bedsides. And it's so solemn. The priests and monks, the psalms, the prayers, the churches themselves. Solemn and cold and joyless. Frightening, almost.' Her hands fell back in her lap.

'You must hate us,' I said, running my fingers through her thick, dark hair, glossy in the firelight.

'No. At least not now. Of course I did at first, but then I realised that there was someone I hated far more . . . and that made me understand that the Gentiles in our town were just ordinary, frightened, ignorant people who went along with each other because they didn't dare do otherwise – like that woman in the inn, I suppose. But they weren't evil . . .'

'And the wells – were they poisoned?'

Her hands flew out again. 'I doubt it, and even if they were, why should we have done it? We had a good life. There was nothing, *nothing* for any Jew to gain by doing that.'

She fell silent and I found myself thinking how much more approachable her god sounded than mine – an especially blasphemous thought, no doubt, given our present lodgings, but one I could not deter. For that matter, was I convinced that my god really was mine? Or was it more a question, as Ellen had said at the shrine, of taking your blessings wherever they were offered? On an impulse, I made a prayer to Ellen's god, whoever he was, thanking him for bringing her to me, and asking him to look kindly on us.

Without moving, Ellen began to sing. The language was quite unfamiliar, a curious combination of the guttural and the sensuous. At first it called to mind the tearing of sackcloth and the droning of bees, but as the music took hold of me I became aware that here was a beauty, a richness, an earthiness far greater than anything I had ever heard in my own tongue.

She sang softly, for fear of waking Roland, and at first the tone of the song was slow and sorrowful, the melody plaintive. But although the minor key persisted, the rhythm gathered pace to lift the mood of melancholy, creating a sense of urgency and a vibrant pulse, surging round and round, full of optimism, vitality and a hint of defiance. I found myself rocking from side to side and drumming my fingers in time.

The song ended and she smiled up at me, flushed and a little breathless. She reached back with one hand to take mine.

'It's about a young man and woman who love one another. He has to go away into the mountains to gather his father's sheep and they worry that their love will dim during their separation. "Why don't we ask Yahweh" – that's what we call God – "to take care of us?" she says. So they say a prayer together, asking Him to bless them. The young man goes off and when he returns, much later, they realise that their love has grown stronger than ever.'

I stroked her hand and told her that it was a beautiful song, beautifully sung. I asked her why she had chosen it.

She laughed. 'No special reason. It's one of my favourites. My grandmother used to sing it to us.' She pulled my head down till our lips met and we kissed for a long time, then drew back and lay together by the fire, talking gently. My thoughts strayed briefly to her body, soft and warm beside me, but as we talked on, our clasped hands echoing the meeting of our souls, I began to feel an extraordinary tenderness and contentment which quite transcended physical desire. Only much later, when the fire was nearly out and we rose from the floor, did I think again of the passion with which we had begun the evening, all those hours ago.

We fed the fire and put out all but one of the candles. There were two more mattresses, a little apart from Roland and screened from him by the corner of a table. I could not help noticing how they had been laid side by side.

We lay down fully clothed, pulled the blankets over us and nestled close until sleep took us away to our separate dreams.

Thirteen

oland's condition worsened towards the end of that first night in the monastery. We had slept only for a little while before he woke us with his moaning and tossing. He was drenched with sweat, his skin wax-like and burning hot to touch.

Shortly after dawn we roused the old man, who came and prepared his second potion. A good deal of it went down Roland's chin into his already soaked shirt, but we managed to get some into his gullet and although he retched violently, it stayed down. For an hour or so afterwards, the cough seemed to become looser and his breathing less shallow. But the infection had not relinquished its hold and by mid-morning he panted and rasped again as badly as ever.

The old man would not give him any more of the remedy. It was too strong, he said, and would kill Roland more certainly than the disease if the doses were taken too close together. We would have to wait until the evening before allowing him any more. All we could do was stay with him, keep the fire up, and wipe away the sweat which, he explained, cooled on Roland's skin and chilled him despite the heat of the fever.

He shared a spoonful of broth with us, then went off to pray.

By now Roland's bedding was also drenched. At Ellen's suggestion we moved one of our mattresses up to the fire, rolled him onto it and, covering him bit by bit with a fresh blanket, removed his sodden shirt and breeches. The slim body was pale and smooth-skinned.

'He looks so young . . .' said Ellen.

I nodded. It was still the physique of a boy.

Ellen settled down on the floor beside him and I went to fetch more wood, then lit a second fire in the cobwebbed kitchen range and spread the damp, sweat-soured bedding around it to dry. That done, I went in search of the monk for directions to the village.

I found him, at length, in the cottage. Its dingy single room still possessed three dilapidated pieces of furniture: a wooden pallet, a small table on which stood a pewter candlestick, and a stool. He was seated at the table, polishing a large and beautifully wrought silver crucifix with the hem of his habit. He worked methodically, buffing the metal with a practised hand. He had reached Christ's knees as I came in and below them the silver gleamed in the dim light.

He held it towards me, a faint smile playing amongst the seams of his face.

'Does it shine?'

'It does,' I replied. 'It's beautiful.'

He nodded. 'Our one treasure. I dare not keep it in the church now, so it stays hidden – here in the abbot's dwelling, may he rest in peace.'

'You're very trusting,' I said, '. . . and very generous. I think Roland would be dead by now were it not for you.'

'Generous?' He gave a little laugh. 'No. I merely live by my faith. And trusting? Well . . .' His voice became serious. 'You have suffered, I believe, all three of you . . .'

'Mmm . . .'

'The Lord sends us suffering so that we may learn and grow, not only in our faith, but in ourselves. You are learning, you and the young woman and your sick friend. I know that. And anyway,' he smiled again, 'what have I to fear from young lovers whose thoughts are only for one another?'

'Nothing,' I said, blushing needlessly.

I returned in the early afternoon, tired by the load of provisions I had been carrying, but buoyed up by a piece of good news. A little under a day's walk from the village, I had learnt, there was a small town on a hilltop, whose church steeple could be seen for several miles around. I hurried into the refectory, eager to check this information against the map, only to find that in my absence Roland had become delirious.

It had started an hour or so ago, Ellen explained, as I sat down beside her. She had called the old man again, but he told her it was no more than the fever taking its course and that she should not be alarmed. She appeared tired and strained, nonetheless, and was clearly most relieved to see me. I soon began to understand why. The delirium was an unnerving thing to listen to, not merely because it suggested a mind disconnected from its normal source of reason, nor even because the ramblings were so different from the directness of his utterances in the moments preceding a fit, but because there *was* ultimately a kind of coherence to what he was saying – an underlying theme of immense loneliness.

Fragments of his life in the village spilled out like the reflections from a cracked blade: the children, who had made fun of him; the reeve, who had orphaned him; his parents, who had failed to provide him with a sibling; even his aunt, who had deserted him. We sensed that his rage was directed not so much at them but at the isolation he had endured as a result of their actions, real or imagined. He called for Nebuchadnezzar and another name we did now know, sometimes tenderly, sometimes imperiously – a pet dog or cat, perhaps, which he had never mentioned. Crebanellen were there too. And someone who had no name, but for whom he seemed to yearn desperately.

'Who is it?' Ellen asked faintly.

I shook my head.

'Has he ever been in love?'

'Not to my knowledge. Not so long ago,' the memory made me smile a little, 'he said that perhaps when we were alchemists he'd know what it was like to love a woman – or something like that.'

'He must be imagining someone.' She looked away.

Throughout the afternoon we took turns to sit beside him, mopping away the sweat that poured from his face and neck, and trying from time to time to coax a little water between his lips. We soon gave up attempting to talk against the low, disjointed monologue which continued like the relentless creak and rattle of a wheel.

Some time after dusk the old man administered another dose of the remedy, but would not be drawn when we asked him if it was having any effect.

'I use what is available to me. The Lord will take care of the rest.' He sensed our anxiety and added: 'It has worked in the past, though. I will vouch for that.'

Roland, though by now quite exhausted, remained in the grip of his delirium. I began to fear that he might pass away from the sheer effort of speaking. His voice had become cracked and the tendons at his throat stood up like cords. By midnight Ellen had fallen asleep on the floor by the fire, and an hour or so later I also succumbed, too tired to care any longer whether he lived or not.

I slept fitfully and woke a little before dawn with the impression that there was someone else in the room. I opened my eyes to find the old man kneeling at Roland's side. I thought immediately of the last rites and my heart tripped, then tripped again as I saw, by the flickering stub of the one remaining candle, that Roland was now utterly still and pale as a shroud. But there was an expression on the monk's venerable face which seemed at odds with the administration of extreme unction. As I scrambled from my mattress, he nodded in what appeared to be mild satisfaction and put a finger to his lips. Looking more closely I now saw that Roland was breathing – almost imperceptibly – but breathing nonetheless. The fever spots had gone from his cheeks and his skin, though still wax-like, was dry and cool again.

The old man patted me on the shoulder and whispered: 'The worst is past. He will sleep now. Another day, I would venture. And then he will need to rest. The fever has left him very weak.' He stood up and made his way softly from the room – a figure of great dignity for all his age and blindness.

I looked across at Ellen, deliberating whether to wake her, and decided not to. With the dawn light starting to seep faintly into the room, she looked so lovely in her deep sleep, the long lashes fluttering gently over closed eyes, dark hair fanned across the floor. And the other recumbent figure, pale, almost motionless, but restored to the living – only now did I admit the extent of my relief.

We stayed on at the monastery for several days. As the old man had predicted, Roland remained asleep until halfway through the following morning. Then, when he attempted to rise, his legs buckled beneath him and he lay back panting on the mattress with an expression of pained surprise.

It troubled me to see how gaunt he had become, and how that air of fragility which had so struck me when I had first met him had now returned. But he was as cheerful and curious as his condition permitted, joking about his incapacity once he became used to it, and wishing to know all about the monastery and how we came to be there – of which he claimed no recollection whatsoever.

Ellen resolved that he should be fed, abundantly. It was the way her people showed concern, she explained, and how could she be concerned for someone who resembled a broom-handle? The argument lacked a certain logic, but since I stood to benefit as much as Roland, I did not complain. A constant and appetising smell of soup wafted from the pot simmering in the fireplace. The old man started to appear more frequently, and although Roland was not the initial attraction, I could tell that he was quickly engaged by Roland's frankness and wit, his natural curiosity and charm.

It was in many ways a peaceful and pleasant interlude, the rigours of the journey temporarily abandoned, cocooned from the outside world in a place where no one else, it seemed, ever came near, eating and idling away the time in conversation with the monk – and Ellen never far away.

But it was not perfect. Close as we were, we refrained instinctively from showing our affection in front of Roland, although – heaven knew – if the old blind monk had guessed what was taking place within a few hours of our arrival, it certainly would not have escaped Roland's sharply observant eye. He knew, anyway. We knew that, and perhaps we held back on account of it. Even so, there was scarcely a moment to be alone together, for the illness seemed to have disrupted Roland's natural rhythm. He would doze off without warning at any time of day or night, sleep for anything from a few minutes to several hours, and wake in a state of almost febrile invigoration, craving food, company and conversation. As a result, a snatched kiss, a few mumbled words of endearment were all we managed, and on the odd occasion when we could have slipped away for a little longer, the same thing prevented us. But it did not prevent me resenting his demands, particularly when he was the recipient of Ellen's attention and I was not.

On the afternoon of the fourth day, however, he took us by surprise with the sudden suggestion that we had been cooped up with him for far too long and should go out to find some fresh air. It was the old, spontaneous Roland, his pale face innocently alive with the pleasure he was proposing. Ellen blushed a little but recovered herself by planting a kiss on his head. Roland rolled his eyes in delight.

We strolled down to the riverbank where we had gathered wood the first night, holding hands and silently savouring one another's company. We stopped and sat down on a fallen tree. Ellen put her arm around me and gently drew my head down onto her shoulder. For a few moments we remained there, then my patience at last deserted me. I pulled her close and hugged her fiercely. She resisted; then her arms encircled me, she raised her lips to mine and we kissed and kissed as if we would devour one another. I reached for her breast and she gave a little moan, her body seeming at first to liquefy at my touch, then tremble with anticipation as she slid a hand inside my shirt. Her lips travelled to my ear where she nibbled gently, murmuring as she did so. I felt for the hem of her skirt and she moaned again. Her hand slipped down to my belly as I began to trace the smooth, delicate curve of her calf, the fuller, softer swell of her thigh and at last brushed with my fingertips the moist, sweet, downy warmth upon which every sense I possessed was now focused.

'Oh Creb, sweet Christopher . . .'

She began to press herself to me, then abruptly pulled away. All the melting sense of expectancy turned to rigid apprehension as I scanned her face. I could see none of the anguish that had been present the first night in the forest. But there was something else there, the evidence of a struggle . . .

She reached up and cupped my face in her hands, a curious mixture of resolution and resignation in her eyes.

'We must wait . . . a little longer . . . until he's quite well again.'

'Is this just him? Nothing else?' The question was out, discordant with anger, before I could stop it. But she gave a small, tense smile of reassurance and nodded.

'God damn him! Why, Ellen? Why?'

'Because . . . he nearly died. We mustn't take advantage of him while he's not properly well.'

'For the love of Christ! This is no advantage. He suggested it.' It was childish, I knew, but I had lost the will to be reasonable.

'My love – he said that we should go out . . .' She reached for my hand but I shook her away. 'Oh, Christopher . . . I know . . . but hear what I have to say, please. Just now, when you touched me, I felt none of the . . . blackness. . . . of before. It was . . .

unlike anything I've ever felt.' She sighed. 'I want you, God knows, I want you . . . every bit as much as you want me. But I suppose . . . because this is so very important . . . and because of everything else . . . I need to be sure that there's no . . . dishonesty, no need to be secret about it, nothing in the way. And until we've told Roland, there still is. Can you see that, my love?'

'So you *will* tell him – once he's well?'

She smiled again and this time I allowed her to take my hand.

'What else can we do? I won't go on pretending. Will you?'

I hesitated for a moment. 'No . . .'

She reached up and stroked my cheek. 'Oh, Christopher, I know what you think about him – and he of you. And I love you both for it. But you can't protect him forever, and in a sense, he already knows – which is why it's still more important that we . . . receive his blessing.' She slipped an arm around my waist and pulled me up. 'Come. We'll walk a while.'

The sun had emerged as we were talking. Now it silvered the winter-bleached wild grasses brushing at our knees as we strolled through the meadow. Beside us, veined with the shadows of naked trees, the river burbled gently along its stony course. Ellen and I were alone together, and although the clamour of my senses had not yet died away, I was ready to concede that she was right about Roland. I was too much in love with her not to. And yet, the child within was still needy . . .

'You love both of us?'

'In different ways, yes. I do.' She laid her head on my shoulder.

'In what way do you love him?'

'I suppose . . . for his innocence, his frailty. He makes me want to gather . . .' She stopped and turned to look at me, a gentle mischief in her eyes. 'Are you jealous?'

'No . . . well . . . yes.' I felt suddenly ridiculous. 'I must sound stupid but . . .'

She stood on tiptoe and flung her arms around my neck. 'You are stupid. I'm yours, Christopher. Truly yours.' She looked up at me with a sudden intensity. 'You need never be jealous. Never, never, never.' Then she laughed and drew me towards the riverbank.

'Let's sit here a little.'

We returned to find that Roland's mood had changed. He had become subdued and silent. I had known him at a low ebb before, but the inward look on his pale, gaunt features was disturbing. It was as if he were watching some part of his soul slip away.

Assuming our absence to be the cause, I immediately felt guilty, and glad also that we had not fully taken advantage of it. Ellen, I imagined, felt the same, for she sat down beside him straight away and tried to coax him out of himself.

He brightened a little as evening drew in and we lit the candles, enough, when pressed, to assure us that it was nothing to do with having been left alone; not enough to tell us what it was that was really troubling him – and not enough to lift the sense of oppression in the room.

As Ellen began preparing supper, my own frustration returned and, with it, the familiar resentment. Far from being able to shrug off the mood, I felt it more acutely as I imagined Roland actively seeking every possible opportunity to solicit her sympathy, whilst also covertly following her every movement with his eyes.

After we had eaten I made an excuse and went outside again, but a stumbling tour of the monastery in pitch darkness did nothing to improve my temper. Returning to the refectory, I went straight to my mattress and left the two of them talking by the fire. This time Ellen ignored me, and although a part of me acknowledged that I deserved to be ignored, another part seemed, for the second time that day, to have reverted to childhood. I lay there for a long time, half-angry, half-repentant, as I listened to the murmur of their conversation and willed Ellen to come to bed so that I could at least reach across the gap separating our mattresses and squeeze her hand in the darkness. But she remained with Roland by the fire and eventually I fell asleep.

I do not know how long I slept, but when I awoke it was still dark. There was a draught creeping under my blanket. I rolled over to see if a shutter had fallen open and noticed that Ellen was not in her bed. Still too drowsy to wonder where she might be, I rose, intending to put more wood on the fire – then saw that Roland's bed was also empty.

A few moments later I was starkly awake and standing unseen in the shadows at the door of the kitchen. A single candle burned in the centre of the floor beside the mattress which had been dragged away from its drying-place at the fire. On the mattress lay

Roland, pale and naked, a look of rapturous adoration on his upturned face as Ellen, also naked, straddled him, her skin glistening in the candlelight, her breasts swinging gently as she rocked from side to side, and in the curious look in her eyes and faint smile on her lips, the suggestion of some subtle gratification which she alone understood. For an instant it seemed as if they were engaged in an act of worship, some hallowed ritual with the power, the beauty even, to transfix the inadvertent observer. When, a moment later, I turned away, I felt compelled to do so on tiptoe.

I remember nothing about the ensuing hours except a cold, grey dawn creeping up over the rolling hills to the right of the track along which I trudged. I had not the first idea where I was, nor where I was going. The only certainty, a dull nugget in the back of my brain, was that I could never return to the monastery.

I walked all that day, as blind to my surroundings as the old monk. A regular stride served to remind me that my body still functioned, even if my heart and mind had seized, and it blunted the pain a little – but only a little. The pieces of glass were there, just as Ellen had described them. It was well past dusk when I stopped, and then only because I was too exhausted to go any further. Wrapping myself in the one blanket I had brought with me, I more or less fell down behind a hedge.

I was off again before first light, marching savagely and mind-lessly until a rumbling stomach at last forced me to consider what I was doing. I sat down at the edge of the path to take stock and regretted it immediately. Every way my mind turned brought me up against something I had no wish to think about. I had little choice but to persevere though, and in the end it became quite clear that there was only one thing I could do. I had my satchel with the map in it, and I had what was left of my portion of the Jew's money, safe in a pouch inside my shirt. So once again the rootless, aimless, miserable Creb was to have his future determined by the damned inaccurate, inconsistent map – and once again he was grateful for it.

Air, fire and water remained – the steeple, the smithy and the well. I reckoned it a journey of five days from the monastery to the well and a final day-and-a-half to the cave. But where was I now? My surroundings left me none the wiser. I did my best to recall what I had noticed, what landmarks I had passed the previous day, but nothing came to mind. And the map, in a situation such as this, was no more use than a two-legged stool. The only certainty was that I had spent last night behind a hedge, which meant civilisation of some sort, and that now I seemed to be straying into deep woodland, more likely to mean the opposite.

I clambered up a large tree and made my precarious reconnaissance from the upper branches. Ahead, the forest dipped and climbed interminably without the faintest trace of habitation. If ever I had needed corroboration of the tale of the travelling squirrel, this was it. I climbed down again and set off back the way I had come, the image of a whole roast pig gaining mouth-watering clarity in my mind's eyes.

I did not get a whole roast pig, nor even part of one. I did very nearly get my stomach filled with something else, though.

I arrived at the small town with the steeple in the late afternoon to discover that it had one inn, closed since the pestilence, and an alehouse stinking of ferrets, whose surly ale-wife refused to inconvenience herself by providing me with food.

It was a curious place altogether, this town: two or three streets down which the wind whistled, decrepit houses appearing to teeter off the side of the hill and a run-down market-place dominated by the church rising spectacularly heavenwards. Why a place like this had such an imposing seat of worship was beyond me. The people did not look particularly God-fearing. Indeed, to my jaundiced eye, they appeared almost imbecilic, every one either wall-eyed, goitred, hunchbacked or slack-jawed. I could scarcely imagine a less prepossessing citizenry. But I did need food, somewhere to sleep, and directions for the forge.

The market, such as it was, was winding down for the day, but a few stragglers hung on, eager to dispose of their remaining wares now unappetisingly spread out on the packed earth at their feet. I bought an undernourished rabbit from one and a couple of wrinkled turnips from another, then realised I had nothing to cook them in. The rabbit could perhaps have been spitted over a fire, but I could not imagine a turnip roasted – anyway, I had become pampered, and grown used to a tasty stew. The image of Ellen was there straightaway, deftly chopping vegetables, stirring the pot . . . I winced and walked briskly across the market-place to where I thought I had noticed a tinker stowing his pack, only to find that he had gone.

I asked a bystander who hawked loudly and discharged a torrent of phlegm before obliging with the information that I would likely find the tinker in the alehouse. So, clutching my rabbit and turnips, I retraced my steps and elbowed my way into the crowded and foul-smelling place. I had taken only a couple of paces through the throng when I felt something give beneath my foot. There was a yelp followed by a low snarl and the excruciating sensation of teeth sinking into my ankle. The place erupted with laughter as I broke into a frantic one-legged jig in my effort to dislodge the small brown dog whose jaws had closed upon me like a gin-trap. I took a feverish swipe at it with the rabbit, but that only seemed to excite it further and it began to bark furiously, despite having its mouth full of my leg.

Leering, sweating, ale-drenched faces swam past me in sickening confusion as I staggered back and forth in agony. At last I wrenched a jug from the hand of a startled bystander and was about to brain the little brute when a wave of cold water crashed down on us both. For a second we froze together in mid-step. Then my assailant let go and shot out of the door with a mortified howl, leaving me suddenly unencumbered and wildly off balance. To a thunder of applause I lurched backwards and landed in the lap of a large, perspiring man who had just sat down again. I heard the leg of his stool snap as we both tumbled to the floor, then a grunt of rage as he clambered to his feet and drew a knife.

The place went silent. I lay there winded and helpless as the fat man scowled at me and slowly brought the tip of the knife to my shirt-front. With his sleeves rolled up and the sweat glistening on his fleshy upper lip, he resembled nothing so much as a butcher deliberating his cut. But just as a look of decision began to harden his flabby features, the ale-wife moved across, empty bucket still in hand, tapped him briskly on the shoulder and said: 'Enough!'

Conversation resumed immediately. The butcher put away his knife without a word and glanced around meekly for another stool.

I climbed to my feet and left as fast as I could. Would there ever be a worse day in my life, I wondered, as the town receded behind me and I limped on into the dusk, drenched, bitten and bruised, but still clutching the rabbit and the turnips.

Some time later I stopped at a tumbledown mill, abandoned beside its silted stream. I thanked God for at least providing me with shelter for the night and immediately went looking for firewood. I would spit the rabbit and devil take the turnips.

It was only then that I realised I was no longer carrying my satchel. I stood in the dark, reluctantly casting my mind back to the alehouse. It could have come off as I had fallen, I supposed, but in that place it was far more likely that someone had relieved me of it during the melée and if that were the case, there was nothing to be gained by returning, nor did I have the slightest wish to. I was therefore now without the map and – more pressingly – without either knife or tinder.

I walked into the mill again, silently cancelling my thanks to God. I sat down on the floor and stared gloomily at my provisions. Uncooked rabbit? No, not even in this famished state. Turnip it would have to be, then.

I bit off a piece and began to chew.

Fourteen

t times over the next few days I felt certain I must be dreaming, that I would shortly wake up again and find myself back at the monastery with Roland and Ellen and the old man. At others, though, I was only too aware that I was alone and that, like it or not, I was going to have to become accustomed to my own company.

For the moment, I did not like it in the least. I felt anxious and exposed. I found myself fretting about finding safe places to sleep before nightfall, starting at unexpected sounds from the woodland around me, worrying about how long my pocketfuls of food would last. It startled me to realise that, much as I had been irritated by the map, I now felt almost dislocated without it. Earth – air – fire – water – tra . . . gold – *pause* – hill – steeple – forge – well – cave. Round and round they went in my head, an alchemical litany swiftly losing all meaning, but to which I clung fiercely for fear of forgetting my direction.

My pace had slowed, partly because of my ankle which, although it had begun to heal well enough, still gave me some pain; partly because I had lost my enthusiasm for this, or indeed any other, journey. Apart from a brief encounter with a pedlar who sold me tinder and a knife and gave me directions for the forge, remarking with a chuckle as he went on his way that I looked as if I could do with shoeing, I saw no one else on the road and spoke to no one in the handful of villages through which I passed.

I found myself recalling a fortune-teller who had come to our village once. The people had flocked around him like flies at a cowpat. It had seemed like nonsense to me at the time. I would have been ten or twelve years old and the future then had seemed a wonderful thing, simply because the next day was always different and secret. Knowing, it had seemed to me then, would have spoilt it all. But the older folk had solemnly taken their turns to be told to prepare for a visit, or to beware of a one-eyed dog, or that there would be a death in the family, or – most improbable of all – that a windfall was at hand. And they had continued to talk about it for days after.

Now I was beginning to understand why. Now I was one of those older folk. The experience of the past weeks had seen to that. An essential part of the present, I had begun to realise, was some understanding of the shape of the future. But how could I seriously address my future when what I wanted above all was to turn time on its head, to obliterate everything that had taken place in the last few days? In the past, it had suited me to imagine myself something of an outsider. The territory and the faces at home had at least been familiar – and keeping them at arm's length had done nothing to diminish their familiarity. But here, where nothing was familiar any longer, where my solitary state made me wonder whether I really existed, company began to seem mightily important.

Even Nebuchadnezzar would have cheered me, I thought.

I reached the forge around noon on the second day. I could hear the clang of hammer on anvil for some minutes before the descent of the switchback forest track led me out into a clearing in a small valley. Two streams converged here, one tumbling over a rocky outcrop behind the blacksmith's cottage, the other winding more gently along the line of the valley floor. A dog sauntered from the cottage as I approached, inspected me briefly, then wandered off again. The clanging stopped, leaving only the splash and burble of water.

The smith, a small wiry man with hair cropped to within a straw's breadth of his scalp, greeted me with suspicion at first, but once I had convinced him that I was harmless and that all I required were directions, he became quite voluble. He had scarcely seen a traveller since the pestilence had passed, let alone one on horseback. Now he was reduced to mending pots and pans and what was the world coming to if a smith who'd learnt his trade good and proper was having to take on tinker's work? Answer him that, if I could. Anyway, a well did I say? A couple of days' walking to the north? Yes, bound to be. Several, shouldn't wonder. Water said to be uncommonly sweet up there amongst the hills – good for the colic too. Couldn't say where though – never been that far. But then again, what with the way things were, might easily have to go looking for stuff to mend up there. Might even come with me. No – on second thoughts – maybe not this time. What was it again – a well? Just keep following the track and ask at the third hamlet, the one with the stepping-stones.

And so I continued, climbing and dipping and, once in a while, emerging on to brackeny expanses of higher ground from which I could now clearly see the hills ahead – not merely a distant horizon, but an awaiting destination, for somewhere there, surely, lay the cave.

Since my departure from the monastery the fine weather had remained, the sunlight creating an illusion, at least, of spring. But now, as I emerged from the forest and began to follow the track across a sweep of moorland, the hills beyond began to disappear under a billow of powdered slate.

By the time I came to the stepping-stone hamlet, squatting in the meagre shelter of a treeless hollow on the edge of the moor, the wind had got up, wrenching from the cloud a squall of driving rain. The stream beyond which the cottages stood quickly swelled to a torrent the colour of well-brewed ale and beneath its frothing surface, the stepping-stones were all but invisible.

There can scarcely be anything more dismal than the dripping overhang of turfed roofs, and that is largely what I remember about the hamlet – that and the dourness, the taciturnity of the rawboned uplanders inhabiting it. Years of existence on the edge of this windswept plateau had taken root in an unforgiving cast to the features and a certain blankness of gaze. I stood amongst the cottages looking hopefully from entrance to entrance and receiving not a flicker of encouragement. No one was going to offer me shelter, that much was evident. I approached the least disapproving-looking – a young woman with two children clamped to her skirts like a pair of snot-ridden gargoyles – and enquired if she knew of a well thereabouts.

She muttered a reply which I could not hear, separated as we were by the small waterfall veiling her doorway. I stepped forward a reluctant pace and asked again. An icy rivulet ran down the back of my neck, then another.

She repeated her reply, this time with a gesture of irritation. Who needs a well here? That was the gist of it.

Were there *any* wells then, anywhere up here?

She simply shrugged.

I relinquished wells and enquired instead about caves. To this she responded with a nod – which could have meant one or plenty – and at that moment I realised I had clean forgotten the single most important feature of that damned map, the little symbol identifying the cave. It had been an animal, of that I was certain. But what? A sheep? A fox? A wolf?

Was there a cave especially connected with animals? I asked.

She looked at me as if I were quite witless and waved both arms vaguely behind her head, suggesting that the whole range of hills was peppered with caves, each one a veritable circus of animals.

The goat, with the habitual perversity of its kind, waited until I was on the brink of abandoning the whole thing before springing into my mind's eye with its beard wagging and testicles wobbling.

'A goat,' I said breathlessly. 'A goat cave, a goat hill . . .?'

She scowled, perhaps in concentration, perhaps because she was finding our exchange as trying as I was, then said: 'Goat Pike, you want. Dunno 'bout no cave, though.'

She turned promptly and disappeared into the stygian interior of the cottage, dragging the children with her.

I stood there for a moment, wondering whether I now had the strength to seek directions from another of those sodden-spirited hill-dwellers, then turned around to find that the remainder had all receded into their cottages too, like so many water-rats vanishing into the riverbank. The young woman had clearly spoken for them all.

I left the hamlet and followed the track up the far side of the hollow and out across the moorland again. The squall had passed and now from a uniformly dreary sky came persistent drizzle. The way the muddy path had been pocked with hoofmarks, the way the heather and mosses and pale, spiky grasses had been nibbled short at its sides made me realise that this must be a drove road, meaning that it was unlikely to lead to the high ground and that sooner or later we would have to part company. But where was the well? And where in all creation was Goat Pike?

There was a sound behind me. I turned to see a cloaked figure on a donkey, a dark, hunched silhouette against the grey sky. It shouted at me and now that I looked towards it, began to wave its arms.

'Creb! Creb!'

Could it be?

No.

It was Roland – and with recognition came the instant, furious release of every feeling I had striven so hard to bury over the last few days. I ran towards him, dragged him off the donkey and hurled him to the ground to roll with him over the heather, punching and jabbing and yelling. For a short while he resisted, then suddenly went limp as a flounder.

'Kill me, Creb. Go on. Kill me, now. I deserve it.'

I gave him a good kick and stood up. He lay there on his back on the sodden ground looking up at me through eyes drowning in guilt, self-abasement and misery. But it was not that which made me pull away. He seemed to have shrunk, from the top down and from the sides in. He was literally half the figure I had left behind at the monastery and I felt as if I could have picked him up with one hand. His skin was the colour of wood-ash, and his eyes had sunk far back into their sockets. Even the fine lines of his face appeared to have degenerated in some imprecise way. He coughed, a long deep hack which trailed away into a sickly bubbling sound.

I turned my back on him and moved away a few paces, too confused by my own feelings to know what to do or say. I stood there, staring into the bleakness of the landscape and then, after what seemed like an age, found myself walking back towards him, helping him to his feet, leading him towards the donkey . . .

'We should find somewhere to get dry,' I said, my voice sounding as if it no longer belonged to me.

He nodded feebly, then let his head fall forward as I tugged at the donkey's halter and we moved off in silence through the drizzle.

It was no great distance to the forest. Once we were there I was able to build a shelter of sorts out of fallen branches and rummage through the undergrowth for enough dry stuff to light a fire.

Roland was in a shocking condition, his eyes darting about like agitated grey insects, his hands shaking so much that he could scarcely get the bread and cheese I gave him to his mouth. But he seemed to wish to talk, to unburden himself, and it suited me then simply to sit and listen.

At first he was only partly coherent in his desperate eagerness to explain that it was all his fault . . . that he had led Ellen on . . . beseeched her to make love to him . . . that I was not to blame her at all . . . that he alone was the villain . . . that he did not even dare ask me to forgive him. It all came out in a relentless torrent of apology, punctuated by spasms of painful, bloody coughing.

Some of it did not convince me. He had not raped her, after all – I knew that much – and so she was hardly blameless, whatever he might wish me to think. Anyway, I was in no state yet to conclude how I felt about who had done what to whom. After a while, I reckoned I had heard enough of this. I asked him what had happened next.

His tone changed and he continued haltingly, as if now he were having to drag his thoughts, his voice, out of some place that did not want to relinquish them.

They had returned to the refectory to discover that I had gone, he said, and almost immediately he had had a fit. He imagined that Ellen must have taken care of him for he remembered nothing until he had woken up, well into the following morning, to discover that she also had gone.

He paused and gave me a bleak look. 'Is this what you want to hear, Creb?'

I nodded grimly.

'I . . . I had no notion what to do, Creb . . . I felt so shameful . . . afraid . . . damned almost. The old man had gone – I didn't know where – and I could hardly endure it . . . being alone . . . with so much guilt . . . I think I must have gone insane . . . I . . . I . . .'

For a moment he seemed unable to continue. Then he turned his head to one side and pulled down the collar of his cloak to reveal a livid weal which ran halfway around his neck. I had never seen such an injury before, but I guessed at once what it was. I shook my head, not knowing what to say.

He glanced at me briefly then looked away again. 'I found some rope . . . tied it to one of the rafters in the refectory. Then I . . . jumped off the big table . . .' he shuddered and went silent for a moment 'but it slipped and I could get my toes on the ground . . . enough to breathe . . . if I kept very still. Then I . . . remained there until the old man came back.'

'How long?'

He shrugged. 'An hour. Two, perhaps.'

I asked quickly: 'And then?'

He hesitated. 'The monk, Creb, he deserves far more than he has, you know. He should be in some great church somewhere. He saved me – not just by cutting me down . . . He's why I am here.'

Shocked and confused, Roland had by this time convinced himself that he had committed every major sin there was, short of murder (and had he subsequently learnt that I had tripped over in the darkness and broken my neck, he would have been easily persuaded of that, too). Fire and brimstone and eternal damnation, he had concluded, were no less than his due deserts. But the old man had taken him in hand and offered him, sympathetically yet firmly, a simple summary of his own position: you, my son, have contrived the sort of situation which only mankind, with all its ridiculous weaknesses, is capable of. The choice is simple, and you will not be the first to be faced by it. Pull yourself together and go forth and try to make amends, or do nothing and pass the rest of your days feeling haunted. He was no soothsayer, he had said, and he could not guarantee that Roland would find me as accommodating as the Almighty might like me to be, but even if I were not he would at least have begun to redeem himself by trying.

Roland had leapt at this quite unexpected prospect of salvation as if it were a bucket appearing at the bottom of the deep, dark well into which he had tumbled. He had set off the following morning, ignoring the old man's parting advice to go easily, and had walked without stopping for a day and a night – until he collapsed on the road and might possibly, he said, have had another fit, although he could not remember anything about it. At that point, anyway, he had come to his senses, and acquired the donkey.

I asked him where he had got it and he gave me an awkward look and mumbled something unconvincing about some people along the way.

I probed him. 'Do you have money left?'

'A little.' He blushed.

'Less than when you set out?' It was cruel, but I could not resist it.

He fiddled with his hands and said nothing.

'You stole it!' I began to laugh. The idea of Roland stealing anything, especially a donkey, seemed suddenly hilarious. 'You stole a donkey!'

He looked up and nodded wanly.

The rain had stopped and I could hear it chomping outside our shelter. Elderly, arthritic and half-starved, it would have made Nebuchadnezzar look like a racing horse.

'What's it called – Methuselah?'

He gave a feeble grin and shook his head. 'No name.'

'And the money?'

'I gave it to the old man.'

'He would never have taken it . . .'

'No. I left it in the dormitory.'

I nodded.

Little by little he was emerging from himself. There was something almost hopeful in his eyes. I was not yet ready to concede, but I was beginning to be glad of it.

'You remembered how the map went?'

'Yes. We'd looked at it often enough.'

'Know where Goat Pike is, then?'

He coughed again, then nodded.

'You do?'

'Yes. There's a dip in the hills – a saddle – you can see from the well . . .' Now he was animated, eager to please me with what he knew. 'If you stand at the saddle, there are three peaks straight ahead. Goat Pike is on the right.'

'And where's this well?'

'Out there – on the moor. It's not far.'

'Why build a well there? They're not short of water.'

'They are if there's a hot summer. The hill streams dry up, or disappear underground or some such thing. And the drovers must be able to water their beasts.'

Out on the moor. Close by. The woman must have known about it all along . . .

'Creb?'

'Yes?'

'Am I . . . am I to come with you?'

I paused for some time before answering. 'I don't know yet.'

Fifteen

rom the well, as Roland had been told, we could see ahead to the saddle, a gentle sinking of the skyline between two rounded hilltops.

The circular stone parapet before us commanded a position of breathtaking bleakness. Somewhere deep beneath the ridge it stood upon there must have bubbled the life-giving force of a spring or an underground stream. But here on the surface, the only evidence of life was a solitary buzzard, planing lazily along the dark smudge of the treeline, a mile or so below us. Everywhere else, the leaden light caused the empty moor to merge indistinguishably with the still emptier hills.

A few hundred paces ahead, the road began to curl away downwards, making for more hospitable territory, as I had suspected it would in due course. I cast another glance at the saddle, listening to my bones protest at the cold and damp and exertion to come. What in the name of God was I to do about keeping Roland warm and dry – myself too, come to that?

He was still sitting on the donkey where it had come to a halt, a hopeful look on its face suggesting it equated the notion of wells with food. But the beast would not get any for the time being, unless it had a taste for bread and cheese, or heather.

'There's no shelter up there,' I said. 'We'd have to sleep out. And while this weather keeps up . . .'

'I'll be all right.'

'No you won't. We should go back to the forest – wait till it clears.'

He shook his head. 'I want to get there. It's no great distance now.'

There was such grim finality in his voice that, much against my better judgment, I conceded. I had to admit, though, that activity – even of this kind – was more appealing than the prospect of another long, awkward huddle under damp branches, each of us thinking thoughts we would rather not.

Our progress, a good deal of it uphill, was pitifully slow. Methuselah seemed incapable of anything more than a senile shuffle. Roland, lulled by the motion into a state of near-trance, was clearly beyond any thought of encouraging him.

Had anyone else been unwise enough to be about under such a dismal sky, they might well have assumed that we were in the last stages of a penance, rather than within a day's march of our crock of gold. It most assuredly felt like a penance. I trudged some distance in front, head down against the drizzle. Roland swayed behind, vacant as a sleepwalker. Methuselah's muzzle drooped closer and closer to the ground.

Had he really persuaded her to make love? And if so, why? There was no question he had known what was taking place between us . . . and malice was one quality he did not possess, I would have wagered my life on that. Still more difficult to understand, though, was how she could have agreed . . . after everything we had expressed for one another. How? Perhaps she really had fallen for him. Perhaps she had just been pretending to me all along. But then she would not have gone off without him . . . unless – I glanced back at him, sagging over Methuselah's neck – unless she had done it purely out of kindness . . . But then why had she been unable to extend the same kindness to me? Perhaps this was also something to do with the time she had spent with her uncle . . . something I would never understand . . . something she might not even understand herself . . . and if it were, did that mean that she was susceptible to everyone who aroused her sympathy . . .? After a while I gave up. As I had already discovered, attempting to make sense of it merely led me into deeper confusion than ever. The only certainty, as real to me as my own heartbeat, was that I still craved her . . . and if there was even the slightest chance she had gone to Roland out of sympathy, then that was a strand of hope to which I could cling.

'Have you any idea where she went?' I had slowed down while I was thinking and Roland was nearly level with me. He looked up vaguely.

'Uh?'

'Where Ellen went – after the monastery?'

'No. I'm sorry, Creb.'

'And the old man?'

'He never mentioned it.'

A few paces further on he asked flatly: 'Will you go and look for her?'

I shrugged. 'Not till we've found the stone.'

Roland nodded.

As if our arrival had triggered it, a wind got up the moment we reached the saddle and the drizzle turned to driving rain. We paused for a moment to take our bearings, but with the light beginning to fail and the cloud low, there was just the impression of a dark and forbidding bulk rising on the far side of another great tract of heathery nothingness.

Directly beneath us, however, where the ground levelled, was the remains of a circular stone wall. I imagined it must have been some sort of shelter or gathering place for sheep. We made our way down to it and found the place on the inside of the wall where we were most protected from the elements, then piled up loose stones into a second small wall and roofed the intervening space with clumps of heather. The resulting shelter was narrow and draughty, but it did keep off the worst of the rain. Lighting a fire, however, proved quite beyond me. We crawled into our tunnel, blocked the ends with more heather and lay back in the gloom. By the time darkness came, Roland already appeared to have fallen asleep; but his constant coughing and restless movements in the cramped and uncomfortable space conspired to keep me awake until well into the small hours. It was one of the most miserable nights I have ever spent.

By morning a transformation had occurred. The sky was cloudless, washed clean and brilliant. Sunlight streamed over the saddle, carpeting the moorland beyond us in velvet and sketching the three peaks so sharply against the dazzling blue that I felt I could almost reach out and touch them.

Stertorous breathing from within the shelter suggested that Roland had survived the night. Methuselah, though, had not been so lucky. The previous evening, too tired and cold to take proper precautions, we had tethered him loosely to a fallen stone inside the wall. The likelihood of him straying far, were he to have slipped the rope, had seemed remote; but in the event he had broken loose, passed through the gap in the wall and wandered thirty or forty paces to where he now lay on his side, a pale grey lump in the heather. It had not been old age or undernourishment that had killed him, however. He had been partially, but quite systematically, eaten.

His throat was gone. The haunches and upper parts of the legs had been stripped to the bone. The belly had been ripped away to the ribs and a mess of blood stained the heather beneath. His eyes stared glassily, his lips were drawn back over large yellow teeth, and one ear was missing.

This was the moment I had dreaded since the journey began, with a dread no less fierce for having been kept to myself. Grey shadows, cohorts of darkness, the devil's foot-soldiers . . . There had been none at home, or if there had, they had stayed in the deepest part of the forest where no one went. But that had been no obstacle to the tellers of blood-curdling fireside tales. As a boy, my imagination had run riot with the patter of stealthy paws, the moonlit flash of a long, slavering jaw, the distant howl. In some dark crevice of my mind, the lupine images had lingered on. Even in the brilliant sunlight, the sight of Methuselah's remains filled me with a deep unease. It had happened so close to where we were sleeping. Had they visited us, sniffing at our shelter in the darkness, and passed on? Or had we unwittingly, fortuitously, set Methuselah as our decoy?

I shivered and walked back to wake Roland. This was not somewhere I wished to stay.

Roland merely shrugged when I told him what had happened. His eyes were red-rimmed and I guessed now that he had slept as little as I. The sunlight made him look more ghastly than ever, hollowing out his features with deep chisel-strokes of shadow.

I gave him an arm to lean on and like a pair of faltering greybeards we set off for Goat Pike.

From the stone circle, Goat Pike had seemed smoother and more cleanly shaped than its neighbours as it rose invitingly into the clear sky. A sharpish climb, perhaps, but not a difficult one.

Several hours later, as we laboured up the lower slopes, I realised to what extent we had been deceived by the distance. The ground broke in an endless succession of steep banks, gulleys, false crests, and was strewn with boulders poking through the heather or concealed in sweeps of waist-high, foxy-smelling bracken.

I found myself beginning to take more and more of Roland's weight, almost dragging him uphill with me. I stopped and sat on a rock to catch my breath. Bereft of my support, Roland stood where I had left him, swaying like a ninepin. He seemed thoroughly dazed now, as if he no longer knew who or where he was. I tugged him towards me and he sat down but his expression remained vacant and I noticed a twitch beneath one of his eyes.

I took a sip from my water-skin, realised it was nearly empty and cursed myself for not having filled it at one of the peaty streams on the moor. I held it out to Roland but he appeared not to register. I tapped his arm and with a great effort he turned towards me, focusing briefly before his gaze dissipated into blankness again. Too weary to try and penetrate his trance, I stoppered the skin and laid it in his lap, then slipped to the ground and rested my back against the rock, feeling the mid-afternoon sun warm on my face. Flies droned in the bracken around me. We should stop here a good while. Let Roland recover – obstinate devil. I was beginning to feel drowsy.

'Creb . . .' His voice had dropped half an octave, as if it were coming to me from the far end of a tunnel. I was on my feet at once, a familiar pinch in my guts.

'Creb . . . Creban . . . Creban . . .' He was having difficulty getting the rest of it out. 'Crebanellen!' It was hoarse, explosive, as if he had emptied his lungs. I began to reach into my pocket for the knife, then stopped. A strange, pale, inward smile had broken across his face. 'Stars in heaven . . .' the smile intensified 'far apart . . . but . . .' he paused 'close . . . at . . . heart . . . yes! . . . Creban . . .'

The smile was erased by a sudden tautening of his features and his jaw clamped tight. I moved closer, ready to hold him as the fit came on, but nothing happened and after a while I realised that by some enormous effort of will he was keeping it at bay. He had begun to strain and grunt, tossing his head and grinding his teeth while with one hand he shredded the fronds from a long stalk of bracken wrapped so tightly round his other hand that it was beginning to cut into his flesh.

I stood by for what seemed like an age until, with a huge sigh, his shoulders slumped and the tension was gone. He shook his head like a dog and looked at me clear-eyed.

'We should be going on, Creb.'

I stared at him for a moment, trapped between his apparent normality and my own awareness that something extraordinary had taken place. Then I helped him to his feet. He set off immediately, without waiting for my arm.

We climbed for another hour, very slowly and in silence. I remained behind him, listening to the rasp of his breathing and wondering what the effort must be costing him. Twice I offered my assistance and twice he refused. But after a while the way became easier. We left the dips and hollows behind and the hill shed its wiry pelt of bracken for a smoother coat of grass. Now the summit was in full view – a long irregular curve, still high above us, filling most of the sky. Squinting upwards, I could see outcrops of naked rock here and there but no cave, nor any sign of goats for that matter. It crossed my mind that they might have gone the same way as Methuselah, but I dismissed the notion hastily.

Without warning, Roland's legs buckled and he slumped to the ground. I knelt down anxiously beside him but he flapped a hand at me and panted: 'Go on . . . go and find it . . . I shall wait here . . . then you can come back and fetch me . . .'

'But it might take a long time . . .' I waved uneasily at the hillside above us.

'No it won't.'

He said it with such conviction that I wondered for a moment whether he already knew where it was.

'Go on, Creb. I shall be all right. But I would prefer not to . . . walk any further . . . than I need.'

The final part of the ascent was the steepest, and when at last I stood panting at the summit, all thought of caves was driven at once from my mind by the magnificence of what lay around me. To the west, the neighbouring peaks rose in sharp silhouette from their tumbling uplands, while everywhere else, bathed in the golden light of late afternoon, the country fell away for mile upon breathtaking mile. Behind was the moor and beyond, the dark ocean of forest spreading interminably to the horizon with, here and there, a palely shimmering island of cultivation. Ahead, a series of plunging valleys, green lakes, the glinting ribbon of a river, a scattering of miniature villages. I could see the whole earth – every crease and fold, every wrinkle and scar on the weathered hide of this giant which slept beneath me. I thought now of the monk and how, had he ever reached his destination, he would surely have felt moved to make a real map, distilling what lay before him, reducing the world to its essence, then recreating it in some way that would have meaning for all the ants that crawled about its surface.

A little distance below me a track wound down and over a grassy shoulder. On impulse I dropped down and followed it. Beyond the shoulder it traversed a long steep slope, covered with a loose, treacherous scree. It was not until I reached the far side that I glanced up and saw the jumble of boulders from which the scree seemed to issue, and above them a ragged cliff with the dark mouth of a cave at its base.

For several moments I stood and gazed, incredulity slowly yielding to elation. The temptation to scramble up there now was almost irresistible, but my conscience would not allow me to. I began to retrace my steps, wondering as I went how Roland would ever manage such a climb.

Slowing a little as I reached the summit, I prepared to yell down the glad news to the small reclining figure below me. But the yell was extinguished by the sudden flutter of a huge, cold moth in the centre of my chest. For a moment I stopped, paralysed, then dashed onwards, slipping and skidding, knowing now that it made no difference, but driven by the need to confirm with my eyes what I already knew in my heart.

The way he was lying, on his back, limbs loose, he could very easily have been in that deep repose which followed a seizure. But the absolute immobility, the open eyes and the faint blueness about his mouth and cheeks, were the indisputable evidence of something else.

I knelt down, gently closing the eyelids and brushing away the hair from his forehead, then began to fuss with his clothes, straightening his breeches, tutting over a hole in his cloak, remonstrating with him for being so untidy, for not taking proper care of himself. There was a little tightness in my head and I had the mild sensation of being not quite on the ground, but I did not stop to question the normality of what I was doing.

Once I had him looking presentable, I sat back and began to talk to him – mainly, I think, about what we were going to find in the cave. I had been talking for a little while before I realised that something was dripping onto the back of one of my hands. I touched my cheek instinctively and found it moist. Reality, at last, overtook me.

Some time later, dazed and exhausted, I lay down on the grass, wrapping us together in his cloak, and fell into a fitful sleep.

A fat yellow moon was creeping over the shoulder of Goat Pike as I struggled back towards consciousness. For a moment I lay still, feeling fogged and disorientated. Then I sat up, and as the cloak fell away, I knew at once where I was and what had happened. I also knew what I was going to do, not only because I could not leave Roland to rot on the hillside, but also because of a sudden obstinate determination to see that he reach the destination he had dreamed of so much. We had set off together. We would arrive together.

I could only have slept for a little while, for although his flesh was beginning to cool, his limbs were still loose and I was able to carry him at first, wrapped tight in his cloak and draped over my shoulder like a sack. He seemed light as a feather, but by the time I reached the summit, my muscles were aching intolerably and I began to stumble at the slightest irregularity underfoot. I laid him down and sank to the grass, vaguely aware that the moonlit valleys below me were now filled with mist. As I sat there, feeling my heartbeat subsiding little by little, I found myself imagining that he would at any moment open his eyes and ask me what we were doing there – and I would reply proudly that I had found the cave.

In time the thought faded. A chill rose from the ground. The shrouded figure at my side lay motionless as marble. I picked him up again and set off down the track, but only a little later, as I stumbled across the scree-covered slope and paused to look up at the dark shadow of the cave mouth, I knew that I could carry him no further. Despairingly, I laid him down once more and slumped to the ground at his side. I do not know how long I sat there before it occurred to me that if I drew the cloak under his arms and tied it around his chest, the other end could be extended into a sling.

With the greatest effort I stirred myself, re-arranged the cloak and began to climb again, now dragging him behind me. His feet scraped dully on the loose stones as we went.

Some time later I stood in the shadows amongst the giant boulders at the foot of the cliff and wept with frustration and rage at the sheer injustice of God and the world He had made. Later still, after an eternity of scrambling and slithering, heaving and panting, I emerged onto the flattish surface of the topmost boulder and, with fingers scraped and bleeding, dragged Roland up after me. Standing there, the floor of the cave was more or less level with my eyes, but although I could heave myself up, I no longer possessed the strength to lift Roland that high, and once I was up there, I could not reach all the way down to him, no matter how far over the edge I leaned. In the end I built him a sort of throne, a pile of smaller stones into which I could wedge him sitting upright. I climbed up, reached down and grasped him by the collar, then hauled.

I am ashamed to say that my hands and wrists had by then become so tired that I let go of him the first time and he slid back onto the throne, then disappeared into the darkness at the foot of the boulder. The rattle of stones echoed eerily across the hillside. But at length I succeeded, and collapsed on the floor of the cave.

Flooded with moonlight, the cave possessed one most distinctive feature. Roughly in the centre, but nearer the mouth than the rear, the natural rock of the floor had been thrust upwards into a rectangular shape resembling a long, low sarcophagus. I glanced at Roland, sprawled beside me. I rose to my feet and dragged him across the floor, then lifted him up and laid him out on the raised stone. I put his clothes in order and folded his now stiffening arms across his chest. There, that was how he should look. Peaceful and dignified. Not unlike the stone figure, a former abbot perhaps, which I had glimpsed through the church door at the monastery.

I felt a little better then. I turned my attention to the cave itself and quickly established that it was empty. Indeed it was almost as if the place had been deliberately cleaned. There was nothing there at all. Not a bone, not a twig, not a bat dropping. Only then did it occur to me that I had no idea what I was looking for.

Roland would have known. But Roland could not tell me now, asleep on his stone.

With a mounting sense of desolation, I ran my hands over the wall, stood on tiptoe and felt the ceiling, squinting perfunctorily into the shadows in the corners. But I could feel nothing loose, see nothing twinkling at me in the moonlight, demanding to be prised out and carried off. It was all uniformly cold, grey and solid as the end of time.

I sat down at Roland's side and began to talk to him:

'There's nothing here, you know. Nothing I can see. What should I do? Bring tools and spend the rest of my life breaking rock? How would I know what it was when I found it? This could even be the wrong cave. Or it could be the right one and the whole thing a wild goose chase. Did you die for a dream, Roland – my friend, my true friend who I never forgave? Did you? And what am I to do without you? Oh Roland . . .'

Dawn was breaking as I said goodbye to him. I left the cave and made my way down towards the mist which filled the valleys below with a deep blanket of uncertainty.

PART THREE

Sixteen

or a long time after, a part of me ceased to function. I could still think and speak, register my whereabouts, conduct the necessary daily exchanges with my fellow men, but I could no longer feel. Whether the disability was temporary or permanent I did not know, nor was I greatly concerned. Nothing mattered and nothing interested me. I swiftly became as self-sufficient as I was numb.

It was this numbness that sustained me through the following months. I found work in one of the valleys beyond Goat Pike. My task, and the irony of this did not escape me, was to load a handcart with stones at the quarry halfway up the hillside and trundle it down to the lake below, where the local manor was being rebuilt. Incensed by the prohibition of alcohol and dice, piously imposed upon them by their lord for their protection against the pestilence, a mob of villagers had one night descended on the manor, driven out the lord, his family and his steward, and thereupon set fire to it. But they had not just thrown a torch or two through the doors: that would have been mere truculence, lacking the proper refinement of real vengeance. No, they had done something far more wicked. They had chosen a night of gales and consumed the entire manorial supply of winter fuel in building a monstrous bonfire around the walls. Only then had they applied the torch and spent the remaining hours of darkness throwing on more fuel. The heat had been so intense that in the lower parts of the building the masonry had vitrified, leaving the place completely beyond repair, yet almost impossible to demolish. But demolished, eventually, it had been – by the villagers. And now that the reconstruction was under way they had been moved in shackles up the hillside where they were discovering the further delights of quarrying.

It became a particularly hot summer and the work was back-breaking, worse even than salt-gathering. It occurred to me that mules or horses would have been better fitted to the task than humans, but I soon learnt that the whole region had been cut off during the previous winter's snowfalls, and that the harsh weather, combined with a virulent attack of pestilence in the nearest town, had resulted in some unorthodox ingredients finding their way into the local cooking pots. Now there was not a pack-animal to be had in the region and the cost of importing from beyond, where their value had soared for similar reasons, was considerably greater than the cost of two-legged transport.

All of which suited me very well. I awoke every morning craving exhaustion. I longed to sweat, to feel my muscles crack and cry out so that come the evening I would fall instantly and dreamlessly asleep. And I succeeded. There was even a hint of grudging admiration from my ox-like fellow stone-trundlers. But I was beyond their friendship, sinking onto my straw mattress in our rank sleeping hut as soon as we had eaten and before the nightly round of dice and ribaldry began.

My one brief human contact was with a middle-aged carpenter who taught me how to swim. I watched him for several evenings, fascinated by the way he could glide through the still water of the lake, a great goose-skein of ripples spreading out behind him. It was something which, for reasons I did not understand, I found myself eager to be able to do.

At length I asked him to part with his secret and after a startled look had slid across his face he shrugged and agreed. But I kept even him at a distance. I hardly spoke during our lesson, thanked him politely afterwards and avoided his company altogether once I learnt the knack of it; which I regret now, because he had a good sense of humour and was extremely patient – I was not the most apt of pupils – and he must have thought me a strange young man indeed.

From then on I swam whenever I could, not because it gave me pleasure, but because it was another way of exhausting myself, because in some curious way I felt I was being cleansed, and because the unrestricted motion of my body through this unfamiliar element seemed to echo the inner void in which I floated and where, for the time being, I very much wished to remain.

One evening, towards the end of summer, I stretched out on my mattress as usual and found myself unable to sleep. A thunderstorm had been gathering since the afternoon. The air was heavy and humid and I could feel the sweat trickling down my chest. I lay there, shifting uncomfortably in the heat, half-listening to the conversation at the far end of the hut.

Someone was telling a rambling tale about a feud with a neighbour over a dog which had the habit of burying its bones beneath the neighbour's hen-coop. After the neighbour's protests had gone ignored for some months, the constant excavation finally proved too much. The coop's foundations gave way one night, the whole thing toppled onto its side, the door sprang open and a fox got in and dined extravagantly on the inhabitants.

At this point, complicated animosities involving further neighbours, their animals and relatives had come into play. The feud grew rapidly out of hand, and the narrator had been driven from his village and forced to take refuge in a cave.

How it all ended, I never knew, because amidst the general laughter someone else interrupted to relate solemnly and insistently how he too had spent a night in a cave. Only a few months ago, as it happened. When he was on his way here, in fact. Not that he'd been on the run, mind. Just run out of daylight. Right odd cave it was, though. Clean as a whistle. Nothing in it but a big stone slab like a coffin in the middle of the floor.

'Up Goat Pike, was it?' asked a local man.

The traveller shrugged, at which the local man began to describe the distinctive trio of peaks at the edge of the moor. As the traveller nodded in recognition, the local man went on to explain that the cave was famous as a hermitage, known throughout the country. Each hermit spent a few weeks at a time there, he declared with authority, always leaving the place clean for the next one to come along.

The traveller shook his head thoughtfully. 'Must be arse-bitin' in winter . . .'

I rose and went down to the lake in the dusk. Distant rumbling suggested that the thunderstorm had passed us by. The sky was clearer now, but the air still felt oppressive and there was an oily stillness to the water.

Was that what my monk had been in search of? A hermitage? Nothing more than a secluded place and time to pray, or think – peace, in fact? Was that what alchemy, or being a philosopher, really meant? Being peaceful? If it was, perhaps we *had* found what we were looking for. Roland at least had found it, there on the stone. The eternal variety. And what about me? Was I at peace? Yes, in a curious way, contained and somewhat fragile, I supposed I was – or had been until now. But what part did lead and gold and fantastic wealth have to play? Had Roland simply invented it all?

I could feel a tightening at my temples. I was thinking too much. Or perhaps it was just the vestiges of thunder still hanging in the air. I stripped off and waded into the lake, swam out a little, then turned on my back and floated. With the silken sensation of the water, the night sky cavernous above me, I began to feel myself gratefully sinking back into my void again. The past was gone, irrelevant.

The next day I took my place as usual among the perspiring swarm of half-naked, nut-brown bodies and laboured mindlessly once more through the haze of heat and dust, assailed by the sour smells of exertion, the incessant ring of chisels and rasp of saws, the creaking of pulleys, the clatter of feet along wooden scaffolding.

But come the end of the day, as I sat down on the little shingle beach and began to unlace my boots, I found myself strangely overcome by the beauty of my surroundings: the sunlit hillsides above me, perfectly reflected in the unbroken water ahead; the evening conversation of the waterfowl echoing out from the slim crescent of reeds at the lake's nearest end; the water itself silvery-brown and clear down to the shimmering, weed-covered stones on the bed and the fish that swam amongst them.

I went to my mattress that night still savouring the image – a small enough pleasure, but the first I had recognised in a very long time.

The master mason was a large man with a large nose, a large belly and enormously large hands. From head to toe he was grey, uniformly coated with a fine film of dust which clung to him at all times except when it rained; then it slithered into streaks and patchy cakes through which a second, darker film became visible. For he was also enormously hairy.

Lumbering about the site, he called to mind a druid stone which had recently, and much against its better judgement, permitted itself to be brought to life. But his dust-encrusted eyes were constantly creased in amusement, as if in the obdurate matter which gave him his livelihood, he had discovered some immense joke.

Although he was one of the very few people for whom I had spared more than a glance, I imagined that his elevated position – and formidable presence – placed us beyond the possibility of any acquaintance whatsoever. And so my jaw dropped when one afternoon he beckoned me away from my cart and informed me that he'd had his eye on me for some time and now wanted me to replace an apprentice who'd contrived to slice off a couple of fingers with a chisel.

'No good to me with less than five on each hand. And no good to himself if he don't take proper care with his tools. Sent him packin' back home. He'll be all right. He can learn milkin'.'

His voice was as large as the rest of him and he spoke gruffly. But the eyes twinkled away at their joke and I suspected he had more sympathy for the young amputee than he was prepared to admit. He drew a long, deep breath then exhaled slowly and a stream of dust motes issued from the whiskery caverns of his nostrils. 'So . . . what do you say?'

I was still too startled to know what to say. I began to mumble, but he waved a hand and stood back patiently as I grappled with the implications of his offer. An end to the dreary, weary round of stone-trundling. Now there was a fine thought. And the chance to learn a real craft. Better still. And a move to more salubrious quarters, the mason's lodge – a nice name for what was really only another hut, but a bigger and more comfortable hut where, I had heard, the bill of fare was somewhat more appetising also. But an apprenticeship. That, as far as I understood it, meant working for him for nothing, bound to him body and soul until he deemed me fit to become a wage-earning journeyman. Until recently I had not given money a second thought; now I was beginning to notice ordinary things once more and the little that was left of Roland's and my reserve, together with the wage I had been drawing as a stone-trundler, had begun to lend a satisfyingly solid clink to my money pouch. I hesitated.

'Creb, isn't it?' My crown came level with his teeth.

I nodded.

'Well, Creb – you thinkin' 'bout wages?'

I nodded again.

He shook his head gravely. 'Root of all evil, my lad.' Then he grinned and slapped me on the back. 'But we like 'em all the same, don't we! Now, you listen to me. I don't know what brought you here and I'm not about to ask – but I do know you've got more in you than what humphin' stones calls for. Been watchin' you, see. Oh, you don't say much, but you work hard and there's plenty goin' on up there,' he tapped his forehead. 'Weren't ready for me to start with, you weren't, but you are now – I know that – and so does whoever takes care of these things. Taken care of 'em now all right – with a chisel. One thing you learn in this craft – there's no hurry about nowt. Things only happen when they're good and ready, not before.

'Now, you work for me and you'll learn all there is to learn about masonry. What's more you'll learn it quick – part because o' me, part because o' you. We'll make this . . . informal – there's a nice word – informal – this apprenticeship. No indentures, nowt writ down. It'll last just till this manor's built. No, I won't pay you, but you'll get fed and bedded and when you leave here you'll earn a better wage than you ever did with that cart o' yours.'

He cocked his head on one side and looked down on me with mild indulgence. 'You with me, young Creb?'

'Y-e-s,' I said, feeling a little overpowered.

His eyebrows beetled. 'Is that y-e-s, or yes?'

'Yes.'

'Grand!'

My hand disappeared into his large grey, hairy paw and I became a mason's apprentice.

By the time the leaves were off the trees and raw, damp winds were whipping the spume from the lake as they raced up the valley, I had been at the manor for more than half a year.

Day by day the building was inching its way up from the ground. Massive transverse beams of oak had recently been mortised into the masonry to support the upper floors, and now carpenters swarmed precariously along their perches, protected from the rain by a temporary roof of woven rushes held aloft on long poles. Meanwhile the walls were rising around them and a spiral stone staircase gained a new step every couple of days. Soon the ladders would be redundant.

Outside, a long, low shelter had been put up against the wall of the building facing up the valley, out of the wind. Here, enveloped in a cloud of fine grey dust, we masons busied ourselves with the hillside, most of which, as I well knew, arrived in crude lumps, just small enough to be manhandled down from the quarry. Occasionally, when a lintel or something still larger was called for, the requisite slab was loaded onto a wooden sled and lowered down the hill with ropes, to the extreme peril of all involved.

Under the master mason's watchful eye, my apprenticeship progressed swiftly. First I 'learnt my stone', examining everything that came down from the quarry for cracks, seams of flint or other impurities, and then sorting it by size. Satisfied that I was not going to allow anything through that would undo the whole edifice, he promoted me to 'shaping' – knocking off the protuber- ances until the lump was reduced to something approximating its ultimately intended form. I learnt quickly how to hold the chisel steady so that the blade did not slip and the mallet blow did not sting my hands; a little less quickly how to apply the correct amount of force, so that the right part detached itself and the stone did not shatter. He seemed pleased enough anyway. Now, to the intense envy of some of the other learners and shapers who were either so inept or so ill-favoured that their prospects for advancement had long since withered under their master's forbid- ding gaze, I found myself moving on again.

'Fining down', it seemed, was the first stage of the craft proper. It called for a good eye and firm hand, as straight edges and right angles were conjured from the shapers' rough offerings to produce an unadorned block that would sit snugly with its neighbours. The work absorbed me immediately, and after my clumsy attempts at carpentry with my father, I was surprised to find that I had some aptitude for it. There was a certain satisfaction in the knowlege that the cleanly shaped stone that left my hands would go straight into the wall of the manor and remain there – barring further arson – for who knew how many hundreds of years. And I found myself recalling with amusement my sense of enormity upon first seeing Porker's castle, as I now began to understand that the mind can encompass any task once its parts have been reduced to a digestible size.

As the first snowfalls came and we moved in close around the braziers in the masons' shed, blowing constantly on our fingers to keep them from losing their grip on the cold steel implements, I found myself beginning to reach a sort of contentment quite new to me. There was something almost meditative about the work: the deliberate chipping, the pauses to sight down an edge, the rhythm of my own chisel strokes, the smooth, cool feel of the stone under my fingers. I felt purposeful but unhurried. And I was aware of a comfortable absence – as if for months I had been gripping something in my hand, white-knuckled with the effort not to let it escape, only to find, as I eventually uncurled my fingers, that there was no longer anything there.

I had already re-admitted the outer world. I now found I could also reflect again. My brother, my parents, Roland and his parents, they had all gone, and while I could not begin to comprehend the grand scheme of which their leaving was a part, it did not seem to matter so much any longer. In varying degrees they continued to inhabit my life as much as the master mason or the chain-gang or the apprentices, and at the moments I missed them it was with a healthy clutch at the heart – a shiver, almost pleasing in its own way, which was gone again long before it could invade my mind.

Then there was Ellen. She alone had failed to settle. Sometimes, wherever it was that she lurked within me, I would feel her stir. Depending on my mood, the stirring – and the thoughts that followed – could be pleasurable or painful. But even that flame seemed to have lost its intensity, so I meandered comfortably along, thinking no further ahead than the completion of the manor.

Meanwhile, I made another acquaintance. There was a clerk of the works – a sharp-featured little fellow who sat in a small hut by the entrance to the site. As each new load of materials arrived, he would dart out and squint intently at the contents of the cart or wagon, one eye twitching as he performed furious calculations on his fingers, then vanish back inside again to scribble in a book.

When not on clerical sentry duty, the ink-rat, as he was known, was forever scurrying about the site, quill in hand, squinting and calculating. He was also present on wages day, assiduously record-ing every transaction as the steward – who was as unlike Roland's father as it was possible to imagine – doled out the pennies with the dour expression of a man who would rather have parted with his toenails.

It was little more than a week after the idea first came to me that I found the opportunity to beard the ink-rat in his sentry hut. If the carpenter had been surprised at my desire to swim, the scribe was thunderstruck when I enquired if he had any writing tools to spare. He gaped at me for a moment, his eye twitching ferociously, then a look of condescension settled slowly on his face. I knew what he was thinking: writing is it, you chisel-wielding clod? And pray where's that going to get you, covered with dust and your hair sticking up like All Hallow's Eve? Get out of here – I doubt you can even read.

But he did not get the chance to say any of it because I had my speech well rehearsed: it wasn't strictly writing that I wanted them for, but I was coming along nicely with my apprenticeship – my master was pleased with me – and I'd been watching the carvers when I could and had begun to get these ideas, see – and now they wouldn't stop – just pouring into my head they were, and if I didn't find some way to save them as they came, I'd forget them all and – well – I'd tried scratching them on bark with a knife, but I couldn't really capture the detail and I'd been watching him lately and thinking that writing seemed just the thing – and – I'd be ever so grateful if he had an old quill and a little bit of ink and maybe even some scraps of parchment . . .

He peered at me, blinking rapidly as he tried to fathom all this. It was an unlikely story, I had to admit, but I had reckoned it would not sound half as unlikely as the truth. I stood there, trying to look eager, appreciative and deferential.

At length he turned, without speaking, and took a shabby satchel off the wall, rootled inside it for a moment and held up a frayed goose quill. Then he reached beneath his deak, produced an old ledger and tore from it the last two pages upon which there were only a few jottings. He handed them to me and tapped the horn on top of his desk.

'Ink. Can spare you a little, but got nothing to put it in. You'll have to get your own horn.'

I began to thank him but he fixed me with a sceptical grin, more rodent-like than ever, and poked me in the chest with one bony finger. 'I want to see these – drawings.'

From then on I was aware of him eyeing me every time we passed one another. He did not say anything, but there was an arch look on his face, and I knew that for reasons of pride – or obstinacy – I was now committed to some kind of tacit wager.

For someone accustomed to the dead weight of a mason's chisel, a quill is an instrument of almost unnatural delicacy and refinement. My first clumsy attempts at draughtsmanship left my fingers black, the quill blunt and the surface of the parchment covered with blotches and stains and wavering lines of wildly irregular width which, scrape with my knife as I might, I was unable to erase completely. But gradually, I became more skilful; although my subject-matter was limited by how far I could climb the hill in deep snow, and how long I could stand the cold once I was up there, by Christmas my efforts had improved quite noticeably. (My inkhorn, I should say, came from the head of a deer, begged from the cooks, then painstakingly whittled out and stoppered, some-what ineffectively, with a plug of wood – thus accounting for the increasingly mottled appearance of the small sack in which I now kept my few possessions.)

As I practised my new craft I was pleased to discover that the snow greatly simplified the landscape, reducing it to black and white and obliterating most of the subtleties of contour. This had the effect of focusing the eye only on essential detail and, in turn, made the business of distance and scale a great deal easier to manage. The inky lines on the pale, greyish parchment were beginning to offer a reasonably precise echo of what I actually saw – in form as well as in colour.

I went to the ink-rat's hut one cold, clear winter's afternoon and laid my parchment down on his desk. He squinted at it, turned it around, cocked his head and squinted again.

'Your master seen this?'

I shook my head.

'Just as well, if you ask me.' He made no attempt to conceal his scorn.

'It's – not quite what you thought it was going to be.'

'You can say that again.'

'Do you know what it is?'

He looked me in the eye. 'No.'

'It's a map – of this valley. Look. Here's the manor . . .' I was trying to contain myself '. . . here's the track. Here's the lake . . . the quarry . . . the hill . . .'

His eyes widened for a moment, then narrowed once more as he looked at it again and began to trace along the lines with his finger, naming the various features as he came to them and nodding to himself.

'Well,' he said at length, expelling air through his teeth. 'A map. Can't say I ever saw one before. Pilgrims use them, don't they – to find their way?'

I nodded jubilantly.

'Not much for pilgrims up here, though . . . What d'you want to make a map for?'

I could not expect him to grasp the deep and curious urge which had been with me now for the best part of a year. I scarcely understood it myself. I shrugged.

'Caught my fancy. See if I could do it . . .'

'Caught your fancy, eh?' He shook his head. 'Well – seems you did it, anyway . . .'

I thanked him again for the quill and parchment and ink. As I walked away I could hear him talking to himself.

'I'll be damned . . . a map . . . he made a map . . .'

He sounded almost appreciative.

Seventeen

inter blew itself out in a succession of gales. Under the pale, scrubbed heavens of early spring, the roof-timbers began to go up on the manor. Stark and skeletal in the still-leafless valley, it looked more than ever like some huge, partially-flayed carcass. Very shortly the basic masonry would be complete and only the work of embellishment remain. My quasi-apprenticeship was therefore nearing its end and so, like a fledgling, I would soon be ejected.

Throughout the rump of the cold season, as I squatted with my chisel and stone in the draughty mason's shed, or climbed the sodden sides of the valley in search of new vantage points, I had found myself thinking increasingly about life beyond the valley, but still only in the most abstract sense, as something vague and formless which did not demand to be addressed. Now, however, I could feel stirrings of uncertainty and apprehension.

One evening the master mason laid an avuncular paw on my shoulder and drew me aside.

'I'm pleased with you, Creb.' I noticed he no longer called me 'young Creb'.

'Told you you'd pick it up quick, didn't I?'

I nodded.

'Fastest apprenticeship I ever seen – in more ways'n one.' He winked broadly. 'Don't go gettin' bigheaded, mind. Ain't no genius – just know how to work hard an' use your nous.' He paused and eyed me ruminatively down his large, dusty nose.

'So what you goin' to do now then, eh?'

I shrugged.

'Move on? Find more work someplace else?'

'Most likely.'

He paused again, then: 'How'd you like to stay on here?'

'As what?'

'A carver.'

It was quite unexpected. I felt myself blush at the compliment. 'Could I . . .?'

'Sure. With a bit o' practice.' He was twinkling at me now. 'Heard you can draw. That's a good start. Anyway, I like havin' you around.'

Was there any reason why not? I opened my mouth to accept, only to hear myself saying: 'Thank you. But I've been here long enough. Time to get going.'

He stopped twinkling – or did he? – and gave me a long look.

'But you got nothin' to go to. Said so y'self.'

'I'll find something.'

'How d'you know? Might be no work elsewhere . . .'

'Do something else, then.'

'Like what – humphin' stones again?'

'If I need to.'

He rubbed his chin. 'Is it a girl?'

'Wish it was.' I laughed awkwardly.

He paused. 'So your mind's made up then?'

'Yes. It is.' I was startled by the force of my own conviction.

Now he smiled, almost wistfully, and nodded. 'Ah well. Didn't think you would in the first place. Still, no harm askin'. So . . . you'll be off in a couple of weeks then. If you ever change your mind, come an' find me. Wouldn't regret it.'

'I know,' I said sincerely. 'Thank you.'

And that was it. I had not so much been nudged from the nest as launched myself from it. I found myself breathless, a little shaken even. But there was a restlessness there – had been for some time – and now that I had acknowledged it, I felt quite different, aware of a pleasing sense of anticipation which almost, but not completely, drove away my apprehension. Powerful as this stirring was though, I could not identify it.

In the end I attributed it, somewhat lamely, to spring.

I left the valley on a day of sudden gusts and scudding clouds. My plan, if it merited the description at all, was rudimentary and rooted in nothing more than the newly discovered urge to indulge myself after all the months of hard work: I would keep moving for a while, see the countryside and amuse myself by making some maps along the way; then, when my money ran out – in two, maybe three, months' time – I would find work.

I paused to let a file of scrawny sheep amble across the path in front of me and cast a last glance over my shoulder at the receding manor, now complete with roof. It occurred to me how very different this departure was from the last one I had made. This was of my own choosing. I had said the few goodbyes I wanted to. And having already discarded so much impedimenta over the last year, wrapping up anything else that had no significance for this journey and stowing it away securely had been the work of moments. My step felt lighter, as if I had abandoned the flat-heeled winter gait and was now rising with each pace on the balls of my feet. Spring. It was an appropriate word.

It took me two days to reach Goat Pike, my first and only destination. I wanted to do or say something – perhaps think or feel something – for Roland. The phrase 'paying my last respects' echoed in my mind as I struggled up the steep, grassy slopes, but it seemed an arid attempt to describe something so utterly personal. It was not only respect, anyway. It was friendship – or love – if there was any difference.

When I reached the goat track I stopped and looked up. Could he still be there? No. The scavengers – or maybe the hermits – would surely have seen to that by now. The mouth of the cave was in full sunlight and I wondered for a moment whether I could actually see the raised stone resting place. Then a shadow slid across the hillside and the cave went dark. I climbed to the top of the scree, scrambled halfway up the boulders and sat down with the cave just above me. That was close enough.

I closed my eyes and pictured him.

The images came readily enough: peering apoplectically into the kitchen as his aunt took her bath; shouting encouragements to Nebuchadnezzar from the raft; blushing at some remark of Ellen's; sitting up on his mattress as he eagerly garnered some new piece of information from the old monk . . . I could clearly recall the timbre of his voice, the knowing grey eyes, the fine pale features shifting with each change of expression. And yet . . . I seemed little moved by all this recollection. I do not know what I had been expecting. A last great tug of the heart, perhaps? Some kind of silent but palpable communion in the empty space between the cave and where I sat? No. There was merely a pleasant glow of affection, spiced with a momentary sadness.

After a while I found my mind drifting away to other things. I stood up and told him, if he was listening, that he was still my truest and greatest friend and that I was deeply sorry he had died without knowing I had forgiven him. Then I clambered down again through the boulders.

It was hungry and thirsty work, climbing Goat Pike. At the summit I stopped to gnaw a mutton bone, presented to me by one of the cooks, and take a swig from my water-skin. I sat down, emptied my sack and spread out the contents on a stone: my home-made inkhorn together with a couple of quills, a corner of pumice, a goat's tooth and a dozen sheets of parchment.

The ratty little clerk, in his off-hand way, had started to take an interest in my map-making – perhaps because the simple lines and empty spaces offered him some sort of relief from the pages of the record-book, tightly packed with spiky words and figures in his own crabbed hand; perhaps for the more mundane reason that he could not bear to see me making such inexpert use of the implements and materials with which he was so familiar. In any event, he had shown me how to smooth the parchment with the pumice stone before starting; how to sharpen the quill properly and maintain a regular flow of ink from it; how to polish the end result with the tooth; and finally, the day before I left, he had presented me with the cloth-wrapped bundle whose contents lay spread before me in the sunlight. Now, like my cold, dead, wandering monk, I was fully equipped. How curious that I should find myself here, a few hundred paces from the destination he had never reached . . .

I took up a sheet of parchment, ruled into eight equal squares. The top two squares already bore a diminutive record of my journey so far and now, while I was high up and could see most of the ground I had covered today, I would make the next entry. It would be a good deal easier than drawing from memory.

I moistened the tip of my quill and scanned the line of the path, winding along the floor of the valley below me. But before I could make even the first tentative tracing, my concentration faltered and I found my eye straying further afield, past the lakes and villages and wooded hillsides to the rumpled line of the horizon, hazy with distance. I stood up and turned slowly around and around, exultant with a new and overwhelming sense of freedom. For a moment I had the feeling that I myself was a blank sheet of parchment upon which I could invent my own future . . .

For the next two days I followed the drove road. The weather was fine for walking and at first I had the company of no one but the larks and hawks and an occasional curlew as we (the road and I, that is) made our way across the moor and began the gradual descent towards lower ground. The emptiness of the landscape forced me to immerse myself in it as I scouted for salient features with which to populate my map. The fresh scents of spring rose from the heather and grasses around me, the air was invigoratingly clear and bright, and I delighted in my solitude.

By the second day I was nearly out of the hills. Villages became more frequent and it began to strike me how shabby and mean-spirited the people appeared, not merely in their dress but in the way they held themselves, in the way they looked at me and, indeed, at each other. They carried with them an air of unwilling-ness, echoed in their evident lack of attention to husbandry and the poor condition of their livestock. It was as if they had lost all sense of worth, and I had to remind myself that I had spent a year in a secluded upland valley, working hard and being decently fed; while down here people were still recovering from the ravages of the pestilence. After all the clean air that had filled my lungs for so long, I thought I could almost taste it, lingering on the lowland breezes; a sensation never more acute than in the deserted villages with their tumbledown husks of empty cottages, the thatch mouldering, the vegetable plots run to seed. I skirted them where possible; where not, I drew my newly-acquired cloak over mouth and nose. Life had departed unpleasantly in these places. I wished to pass through them swiftly and untainted.

A week after leaving the manor I came to the top of a knoll and found myself looking out towards a shallow valley, a mile or two off, in the middle of which stood a feature I would have recognised anywhere in the world. It was the fairy hill. The drove road must have swept me through an enormous half-circle to have brought me back here again.

I stood and gazed at it for a long time, feeling a deep and unaccountable restlessness, an irresistible pressure expanding to my very extremities. Before I could bring my mind to bear on what was taking place, I found myself striding away from the road and off in the direction of the round, grassy mound.

I reached the monastery in the middle of the afternoon. Little had changed, apart from the appearance of a few more holes in the dormitory roof and a good many more weeds in the courtyard.

I spent several anxious minutes searching for the old monk. He was not in the abbot's dwelling, nor anywhere outdoors, and I paced around the buildings calling out, but there was no response. Reluctant to enter the living quarters unless I had to, I tried the church. I opened the door and a long shaft of dusty light penetrated the nave. He was on his knees before the altar, a simple table of oak, long bereft of its altar cloth, but graced now by the silver crucifix, still beautifully polished and glinting defiantly at the cobwebbed gloom and dilapidation around it.

He heard me come in but remained at his prayers. I stood by the door and waited, wondering again at the faith that sustained him here in these dismal surroundings, cold, blind and hungry as he reached the end of his days. Was it faith which had brought me here? No. Only hope. Faith was what was left when even hope had gone – and I did not yet know whether I had the capacity for that.

At length he stood up, slowly and painfully, removed the crucifix from the table and walked towards me. I bade him good day and began to explain who I was but he interrupted me immediately and said, with the glimmer of a smile: 'Ah yes, I know the voice. My young friend who left so suddenly. Come. We will find somewhere a little more comfortable.'

There was a gathering knot in my stomach as I followed him across the yard, into the outbuilding and down the dark corridor, past the kitchen and into the refectory where I guided him to a stool and sat down opposite.

'I brought you some food,' I said, placing bread, cheese and a handful of little sweet cakes in his lap. The topsy-turvy village had been restored to its diurnal routine and the inhabitants, although still a little pasty-looking, now presented all the appearances of normal behaviour. Keeping my distance from the inn, I had managed to buy a few provisions there.

He ran his fingers over them and thanked me, then paused thoughtfully before asking: 'Your friend, did he find you?'

'Yes. He did.'

He smiled and nodded again. 'And now?'

'He . . . died.'

The old man looked momentarily stricken. He made the sign of the cross and I added hastily: 'Not the lung-fever. You cured him of that. We were very grateful. But he had fits.'

'Poor young man. He was so very agitated. I think he found his dilemma more than he could bear.'

'You helped him. He told me.'

He nodded, composure returning. 'The Lord helped him. I am glad.' He tilted his head towards me. 'And now . . . tell me what brings you here?'

I was suddenly tongue-tied. To voice the question which hung on my lips was to make such an irrevocable admission to myself. But it forced its way out: 'Ellen . . . the girl . . . did she . . . say where she would go?'

He smiled, then rose from the stool, motioning me to stay where I was, and made his way out of the room. A minute or two later he reappeared and pressed something into my hand. It was the little polished pebble on the thong she used to wear around her neck.

'She . . . left it behind?'

'In a manner of speaking. She gave it to me – to give to you.'

My heart lurched. 'When? Why . . . I mean . . . did she say anything?'

'She said if you ever came here again I was to give it to you. And if I thought it proper, would I remember you in my prayers – all three. Which I did – and still do.'

'But . . . nothing about where she was going?'

He shook his head. 'She also was very distressed. I think she scarcely knew what she was doing. Certainly not where she would go.' He paused and smiled again. 'But I did learn where she had gone, a little later . . .'

He named a place I had not heard of. It was a long way off, he said, to the east.

'Not where she came from?'

'Perhaps. I would not know.'

'I think not. They . . . burnt all her family – alive – for being Jews.'

I regretted the indiscretion immediately, but he merely sighed and shook his head.

'So much intolerance. And with so little cause. We are all one in the eyes of the Lord. No. I believe you are right. She would not have returned.'

'But how do you know – where she went?' My hands were trembling.

'She sent word. Some months later, it was – I am afraid I cannot recall precisely when – time has very little meaning here, you understand. But it could have been the summer, perhaps. Yes. It was warm, I think. Whenever it was, she found a fellow who was travelling to these parts . . .' he paused and nodded appreciatively to himself '. . . something of an achievement, I believe. There are few people abroad now – following the pestilence – so I am told. We live in sorry times . . . but where was I? The fellow . . . yes, he came and told me of her whereabouts and that she had work as a seamstress.'

'That would have been the good part of a year ago. And no word since?'

'No.'

'So she may still be there . . .'

He smiled reflectively. 'Love is a powerful thing, is it not? I have my love for God and His work. You have your love for a woman. That is beyond my experience – but not, I daresay, beyond my understanding. You should find her. There is much for you to learn from one another.'

I found it hard to reply. My eyes were stinging, my throat was constricted, and a map formed in my mind – of a great country with me one side of it and Ellen the other and something joyous streaming across it.

At last I understood my restlessness, and furthermore, that my own sheet of parchment might not be quite as blank as I had imagined.

I stayed with him that night and slept in the dormitory. Before leaving the next morning, I went into the church with him, placed the crucifix on the oak table, and knelt beside him to pray for Roland's soul.

Eighteen

y first glimpse of the sea came quite unexpectedly. I do not know whether I was more surprised by what I saw or by the circumstances in which I saw it.

I had been climbing a little used forest track for an hour or so, singing and whistling and twirling my stick all the more purposefully as the way grew narrower and the undergrowth more dense. At the summit, the track levelled out, and while I paused to catch my breath, something small and striped shot rapidly in front of me, rustled loudly in the bushes, then trotted out into the open again. It was a wild piglet. Catching sight of me, it flattened its ears and froze. Thoughts of a tender chop invaded my mind. I hurled my stick at it as it sprang to life and scurried squealing down the track.

I walked on and was about to retrieve the stick when there came the sound of something very much larger breaking through the undergrowth behind. I turned to see the piglet's sire approaching at gathering speed. A thundering brown wedge, razor-back bristling with rage and tusks flashing in the filtered sunlight, it was an altogether unnerving sight. Instantly aware that I had neither the speed to outrun the creature, nor the courage to stand my ground and sidestep nonchalantly but niftily at the last moment, I exercised the only remaining option which was to spring, with an agility I did not know I possessed, for the lower branches of the nearest tree.

It was not a dignified performance. As my hands closed on a branch I felt the drawstring go at my waist and a sudden draught around my nether parts as my breeches slipped to my ankles. There was a rending sound as the boar hurtled between my dangling feet, skewed to a halt a few yards further on and shook its head furiously to dislodge the large piece of cloth which flapped from one tusk.

Praying that the branch would hold me, I heaved with all my might; hooked over one elbow, then the other, and heaved again to collapse across it, head and legs down and arse bare to the heavens. Blood pounded at my temples and my vision swam. The boar eyed me and gave a contemptuous snort. It ground one trotter slowly and deliberately into the seat of my breeches, which it had now shed, then began to advance. My perch creaked ominously. I edged towards the trunk and scrambled higher up. Beneath me, my tormentor lifted its snout and sniffed the air awhile. Then it began to circle the tree. For some time it ambled round and round in a leisurely manner as I, still trembling from my narrow squeak, looked down in morbid fascination. At length, however, it appeared to lose interest and wandered off into the undergrowth.

By no means convinced that it was safe yet to descend, I made myself as comfortable as I could, jammed into the fork of a branch a dozen feet above the ground with the trunk hard against my spine, and inspected my surroundings. Although the view consisted largely of foliage, by craning my neck, I was just able to follow the line of the track and beyond it, some way in the distance, I glimpsed a sparkle in the sunlight. I climbed higher up, as far as I dared, and my view suddenly improved. Ahead of me, at the foot of the woods, the ground opened out and shelved gently down for a mile or so before ending abruptly in the sharp line of a cliff-top. And beyond that lay the infinite sea.

I had been aware of it throughout my childhood, of course. The brisk smell, the pull of the tides in the estuary, the sleek dark diving birds sitting on the rocks by the salt pans – these were the manifestations of an invisible neighbour whose presence pervaded our lives. But I had never seen it. From an early age we had shuddered at the lurid tales of pirates and dread sea-creatures and monstrous waves; even the most adventurous of us had been deterred from slipping away and walking the three or four miles of shoreline to the mouth of the estuary. The dense and, so we believed, equally perilous forest was further deterrent to anyone with a mind for the inland route; and curiously, riparian folk though we were, the village had possessed not a single boat. Our lord, the word went, did not consider salt water a fit environment for humankind and took a dim view of water-borne enterprise of any sort. Our lord, in fact, knew very well that boats would have distracted us from the furtherance of his own miserable enterprise; worse still, they would have given us a taste for a kind of freedom which ran contrary to every idea in his head. On land, confined by estuary, forest and escarpment, he had us just where he wanted us and the very notion of the sea, although partly responsible for his fortune, had been something to be discouraged by any means possible.

But here it now was, immense and placid. It reached to the limit of the world, an endless shimmer of silvery blue, dancing with motes of sunlight, and then became the sky. So much water. Where did it come from? What contained it? What if the world moved, even a little, and it slopped, like gravy off a platter, onto the land?

I broke off a piece of branch and lobbed it into the bushes where the boar had gone. There was no response, so I scrambled down from my perch, retrieved the seat of my breeches from where it lay, tattered and muddy in the grass, and set off briskly along the track.

The rigours of the journey had long ago caused me to acquire a certain proficiency with needle and thread. Now, sitting cross-legged on the cliff-top as I stitched my breeches back together, I noticed, some distance out, a small ship slowly following the coast. To my uninitiated eye, it appeared that whoever was at the helm must be either asleep or drunk. Her course seemed most erratic – a continuous and laborious zig-zag involving frequent filling and slackening of her sails, leading everywhere but along the straight line I imagined she should have been pursuing. But the ship's sluggishness was deceptive.

Some way beyond where I sat, a headland thrust itself out into the sea like a knuckled fist. By the time I had completed my tailoring and set off once more, the ship had already rounded it and disappeared from view. Beyond this first headland, at no great distance, lay a second. Her course would have to take her out around that one, I reasoned, and so, quite shortly, I should see her again. But several minutes passed and she failed to appear. Encouraged by the sudden memory of Ellen's wandering flagellants, my mind began to fill with melodramatic thoughts of shipwreck – the vessel sundered on the rocks, the insensible captain flung overboard by the impact and the crew clinging desperately to splintered timbers in the water. I strained for the piteous cries of the drowning but heard only the muted wash of waves and the calling of gulls.

I reached the headland breathless and found myself looking down on a scene as pleasant and peaceful as anything I could have imagined. Directly beneath me lay a little cove. Half a dozen dwellings nestled in the shadow at the foot of the cliff and in front of them, in full sunlight, several small boats were drawn up on the sand. On the far side, at the tip of a low promontory running out through the shallows, the ship rocked gently at her mooring. Voices rose from the knot of villagers who stood on the rocks, exchanging news with the crew who seemed to be preparing to cast off again. Faint smells of pitch and fish piqued the air.

I looked on for a while, wondering what business had brought the vessel to this tiny, hidden place, then caught the sound of other voices from what appeared to be much closer at hand – almost directly beneath me, in fact. I stepped closer to the edge of the cliff, looked down and saw that a crude stairway had been carved from the rock. And now, climbing into view, came a most incongruous procession.

The final stretch of the ascent was so steep that it would have been almost impossible for anyone to look up without toppling backwards. As it was, each member of the procession seemed largely absorbed with his or her next step, and while they puffed and cursed and sweated their way up, I was able to scrutinise them quite blatantly. There were nine of them, and although from a certain amount of breathless banter it was clear they were no strangers to one another, the harder I looked, the harder it became to imagine what could possibly have brought them together.

It was three months since I had left the monastery, a much longer time than the distance warranted, but a certain sequence of events had overtaken me. Before the first month had passed, I ran out of parchment. My maps had by then assumed such importance to my journey and the manner in which I perceived it that I had felt almost unable to continue. I had made for a large town where I heard that replenishment was to be had, and there found myself seduced by what seemed to be an invaluable novelty – a little set of waxed writing tablets, not much bigger than a tinder box, held in a tooled leather slip-case which I could hang from my belt and so have the wherewithal to hand for instant sketches. But the purchase had seriously depleted my funds and I ended up having to find work. As luck would have it, the foundations had recently been laid for a most splendid church, conceived in celebration of the town's deliverance from the pestilence, and I had been readily engaged as a casual labourer. Watching the masons at work as I once again trundled stone, my pride briefly nagged me to reveal my new skill, but thoughts of Ellen had prevailed to keep my mind on the stone-cart and as quick an exit as the wages permitted. It still took several weeks, though.

In any event, I had learnt a great deal during these last three months; more, in the most general sense, than at any other time since first setting out with Roland. As a solitary traveller I had grown used to making the best of whatever company I found myself in. Some instinct for social, and occasionally actual, self-preservation taught me to reveal as little of myself as possible, while at the same time discreetly drawing out my companions until I was certain of the ground beneath me. It was extraordinarily easy. I assumed an interested expression, asked uncontentious questions in an appreciative tone of voice, and most of my interlocutors swiftly opened up like overripe fruit – whilst I became as invisible as a pane of glass. If the majority of my chance acquaintances remembered little of me as a result, the technique at least allowed me to avoid a few potentially unpleasant situations and enter into the spirit of the others to the extent that my preoccupations permitted; and it left me with the gratifying sense of having become a nice judge of the various stations of life.

So now, as this ill-assorted troupe of seafarers laboured up the steps below me, I set myself guessing.

Leading the procession, his florid features drenched with perspiration, was a corpulent and evidently prosperous person. A merchant – that much was likely. But of what? The unusually fine texture of his travelling cloak, now thrown over his shoulder and grasped tightly with one podgy hand as if it would help him haul himself up the hill, suggested wool itself or perhaps woven goods. Behind him came a slightly less portly woman of about the same age, wisps of grey hair falling from beneath a costly looking wimple as she held her head down and her eyes firmly on his heels – whether from uxorial devotion or the need to keep herself going, it was hard to tell. His wife, in any case.

Next, a friar. A tall, scrawny fellow with a cast in one eye and that expression of almost ferocious supplication which suggested he belonged to one of the mendicant orders. The nondescript colour of his habit narrowed the field further: a Franciscan, or Grey Friar. After him, a younger man, not wealthy, but neatly turned out nonetheless, whose clear eye, upright bearing and firm grip of his staff marked him as a person of some military acquaintance. The son of a knight perhaps. Hard behind him, her skirts hitched halfway to her thighs, grimy bodice falling from one shoulder and a trickle of sweat disappearing into a breathtaking cleavage, came a young woman whose less prominent features were presently concealed beneath a stork's nest of blonde hair, held in place by what appeared to be a handful of twigs. It seemed most improbable that she had anything to do with the military fellow, or indeed the type who followed her, a prim and condescending-looking individual of middle-age, wearing an over-elaborate tunic and too much cheap jewellery on his fingers. A craftsman of some sort, maybe in the leather trade – a glover perhaps, and a middle-ranking member of his guild who believed himself worthy of, but would never attain, high office.

I looked on to the last three, intrigued to discover which of them the wench was attached to, but at first glance it might have been any of them. There was a shabbily-dressed elderly man with drooping eyes and a doughy complexion which made me think immediately, but illogically, of bakehouses. There was a country fellow, or so his rustic garb and nut-brown skin told me, who propelled himself awkwardly up the steps with the aid of a stick. A closer look revealed that he was missing most of one foot. A woodsman, I concluded, who had suffered a mishap with his axe. And last, a coarse but pleasant-looking woman of indeterminate age and undistinguished apparel who could well have been the wench's mother, or the wife of either of the two men, or both.

As they began to draw close, I walked off some distance and found a seat on a stone where I composed myself as if having paused there for a rest. A few moments later, the merchant's head appeared over the cliff-top. As he stepped onto the grass I acknowledged him with a wave, at which he nodded, attempted to return the gesture, then rolled his eyes and sank to the ground.

His wife stumbled to the grass at his side, her bosom heaving, as one by one the others appeared and, with the exception of the young military type, collapsed in the extremities of exhaustion. Even he leaned heavily on his staff.

Not wishing to appear too curious, I turned to one of my tablets and began a sketch of the day's journey. But my mind still turned on the reason for their association. They had something more than a sea-voyage in common, I was certain, especially one with such an out-of-the-way destination.

I did not have long to wait for the answer.

The friar was the first to recover his wind. He stood up, a taller and more angular figure still, now that he was not bent to the cliff-face, cleared his throat and said: 'Brother pilgrims, let us praise God for our safe passage.'

The others hauled themselves reluctantly to their knees and a ragged *Te Deum laudamus* ensued. The merchant's mottled, purplish complexion suggested that a second rendering, in thanks for their conquest of the cliff, might have been in order. But the friar now addressed them again, his Adam's apple bobbing nervously: 'Brothers . . . and sisters,' he cast a sharp glance at the wench who was absently scratching her rump as she gazed with patent longing at the young military type, 'fellow pilgrims, our destination is at hand. We will reach the shrine the day after tomorrow. Let us recall that we have undertaken to make our pilgrimage pure in heart and mind. But let us not forget that it is in the last stage of such a journey that our emotions may often overcome us, that we become so – enraptured – at the prospect of laying eyes on the holy saint, that we forget ourselves. Let us especially now conduct ourselves in the most seemly manner so that we shall honour the venerable remains with due piety and sobriety.'

He paused for a moment, then, with a pointed glance at the baker, continued: 'In particular, let no one approach the holy place who has taken strong liquor. Let no one race another to be first there. Let no one write his name or paint his arms on the walls. And let no one use iron tools to prise out a piece of stone for a relic.'

It sounded as if he had made that speech before. But not with reference to this particular shrine, apparently . . .

'Well said, Brother!' exclaimed the glover with a self-satisfied curl of the lip. 'Shall we be on our way?'

There was a general shuffling as the others prepared to move off. But a sudden panic had overtaken the friar. He stared wildly around him, his eyes darting in opposite directions. The shuffling stopped and an expectant hush fell. It was broken by the woodsman who observed laconically: 'He don't know which way to go.'

'Oh, preserve us,' wheezed the merchant's wife.

A look of glee spread across the baker's face and he began to bray like an ass.

The merchant glared at him, then turned to the friar. 'Is this so, Brother?'

The friar nodded.

'You did not think to enquire down there?' He pointed over the cliff.

'I did not.'

'Pray why?'

The friar hesitated, then said hurriedly: 'Because I believed our landfall to have been somewhere else, and had it been so, I would have known our whereabouts – from the instructions I received, that is.'

The merchant looked momentarily puzzled, then his wattles began to quiver.

'That, sir, is gibberish!' He thrust his head forward. 'What precisely do you mean?'

'I mean,' said the friar collecting himself, 'that our worm-begotten captain, upon whom God have mercy, has landed us short – that is to say, not at the place so convenanted between us which, I'd hazard, is some hours' further sailing up the coast and whence I have the most clear directions for the shrine. But this is not that place . . .'

'By the Holy Bones!' exclaimed the merchant, turning his back sharply on the friar.

'An outrage,' muttered the glover. 'The fellow should be strung from a thousand gibbets.'

'He'll have sailed anyway, most likely,' offered the baker, strolling away to peer over the cliff. He turned back with a cheerful nod of confirmation. 'So that's us flummoxed then.'

But the young military type, who had been listening in growing exasperation, now stepped forward and said briskly: 'Come, come. Just look around you. There's a fellow, there.' He pointed towards me and raised his voice. 'You sir! The shrine of Saint Huctred. Can you direct us?'

I put away my tablet, walked over to them, and explained that I was as strange to these parts as they were. The only place I knew of was the town I was heading for. If that was of any assistance, I would be glad to accompany them.

The friar lowered his head as the merchant replied with evident relief that it was the very place recommended to them for lodgings, for the holy shrine was only a few miles beyond and could be visited from the town within the day. He was certain he would speak for his companions in saying that he would be most grateful for my guidance. And he personally would see to it that I dined well when we got there.

It suited me very well, this arrangement, I thought as we moved off. Now that I was so close, I had begun to feel increasingly anxious about the possibility of Ellen not being there, and also, paradoxically, about seeing her if she were.

Company would help keep my mind on other things,

In the course of my journey I had become adept at coaxing comprehensible and accurate directions out of even the most apparently witless individuals. (I still remembered the imprecision of the monk's map with some rancour.) After having been fooled a few times, I had also come to recognise the glib types who would reel off some entirely invented route for the simple satisfaction of appearing knowledgeable or – as happened once or twice – for the malicious pleasure of seeing me heading south when I should have been going north. Meanwhile, I had developed the knack of committing to memory even the most labyrinthine itineraries, and where I remained in any doubt, I now had the materials to record them.

The whereabouts of our destination were a case in point. As we made our way along the coast, I paused to consult my tablet. The next landmark, it seemed, was a lone and very stunted tree on whose bole was a curious deformity resembling the face of an ox, at which point the path turned inland.

The young military type, who had been walking behind me, stopped and asked me politely, but with evident interest, what I was doing. I began to explain but was shortly interrupted as someone else made their way up from behind and tugged at my sleeve. It was the third woman and she was looking at me very strangely.

'Is it . . . are you . . . Creb?'

'Yes.' A dim bell of recognition began to clang in my mind.

'You'll not remember me, but . . . Agnes . . . from the village . . . Hubert, the old mole-catcher . . . his niece . . .'

It was a strange moment: the brief feeling that this was not real flesh and blood, but some spectral visitation from a past I had no great desire to be reminded of made me shiver. But having admitted my identity, courtesy demanded that the encounter run at least some of its natural course. Curiosity also.

She had been a minor player in the pageant of village life, the wife of a weaver whose undisguised contempt for her inability to bear him a child had merely compounded the misery inflicted by a father who was known to detest women in general and his daughter in particular – which was doubtless why she chose to introduce herself by reference to her uncle. My vague recollection was of someone who seemed always to be trying to make herself as inconspicuous as possible. That timidity persisted now, and may have accounted for my failure to recognise her earlier on. But although at first she spoke hesitantly and fidgeted with her hands, as she warmed to her story a certain intelligence began to shine through, and a pleasant expressiveness brightened her face.

The reeve's reign had been short – no more than a few weeks, she told me – and had ended the way it began, as a result of sickness, although somewhat more directly this time. A bloody flux had swept through the village, carrying off more than half the inhabitants, including the reeve himself and the weaver, her husband. God's vengeance for the murder of the steward and the debauch that followed, she added with some satisfaction. But there had been innocent victims too, and it pained her to have to tell me that my father had numbered amongst them. Not my mother, however . . . The village, she continued, had begun to disintegrate as, one by one, the survivors departed in the belief that a better living, and a more congenial atmosphere in which to pursue it, could be found elsewhere. In the end, fewer than a dozen people remained, mostly the elderly. My mother had stayed and there was every reason to believe that she was still alive and well. She said this with a note of encouragement and also, I thought, a hint of admiration.

She herself had joined the migrants and in due course met up with a packman who tempted her with the romance of the travelling life and then, like the previous men of her experience, proceeded to abuse her. She had stayed with him for several months for fear of the alternative, but at length plucked up the courage to give him the slip in some small town in the south-east of the country, and after a few precarious weeks on her own had met the recently-widowed woodsman, to whom she was now wedded. Her luck must have changed, she added with a self-deprecating laugh, for her new husband was a good man through and through. He'd said not a harsh word to her, nor so much as lifted a finger against her, since they'd met.

In any event, she went on, if it hadn't been for the pestilence she wouldn't have met the woodsman and he wouldn't have been widowed and although it didn't seem quite proper to be grateful for such a thing, God must have had a hand in it somewhere, so they'd decided they should make some special act of thanksgiving. God had then intervened further, it seemed, for only a day or two later they'd heard at market that the merchant (who traded in grain, not wool) was arranging a pilgrimage to the shrine of Saint Huctred – a most holy saint who had performed many miracles of healing in his time, and whose name was presently associated with a remedy of mushrooms which, she'd heard for certain, had cured dozens of the pestilence. What could be more fitting?

By that time the merchant had engaged the friar as guide and spiritual counsellor and two others had already pledged themselves – the orphaned nephew of a knight, recently returned from a foreign war, and the cordwainer (I had been right about the leather at least). Shortly after, the thatcher and that daughter of his had completed the party – although what the two of *them* knew about the veneration of saints wouldn't fill a thimble, and she couldn't for the life of her understand what had brought them along. Anyway, to cut the rest of the story short, they'd departed by sea on account of it being quicker, less arduous, and generally safer in these times than travel by land. And the rest, I knew.

I thanked her for the news of my parents and remarked upon the coincidence of our meeting, then told her a little of my adventures since leaving the village, explaining merely that I was hoping to meet someone in the town ahead. After a while the conversation petered out and she withdrew on the pretext of rejoining her husband, leaving me to think about what I had learnt.

My father's death saddened but did not surprise me. I was glad to hear my mother was still alive and promised myself that I would return to see her, sooner rather than later, if it were possible. But while I registered the news with my mind, the village now seemed so remote that it left little impression on my feelings. The much stronger sensation that lingered, as we followed the path away from the sea, was that this chance encounter was somehow significant as an omen for the future.

Nineteen

he pestilence had clearly worked wonders for Saint Huctred's reputation and now the town's innkeepers and tradesmen were merrily reaping the consequences. The place bristled with pilgrims.

The first inn we approached was already filled to overflowing – even the stables had been turned into a dormitory – and we were forced to elbow our way through the crowded streets to a second, then a third and finally a fourth before we were grudgingly shown to an outbuilding. For an exorbitant sum, it was proposed that we should pass the night on a scattering of filthy straw in the company of a number of large and unusually agile rats.

By now the merchant once again appeared to be on the brink of apoplexy while his wife stood rigidly at his side, her eyes bulging with disgust. But the consensus was that we were unlikely to find anything better and so would the merchant, as treasurer, please put his hand in his purse. With very great reluctance, he eventually did so. His wife sat down and began to wail. The cordwainer adopted a pained expression and clapped his hands theatrically to his ears as the rest of us moved in and began gingerly staking our claims to a few feet of rancid straw.

The thatcher, however, had become suddenly galvanised. He strode purposefully into the middle of the room and stood there, one ear cocked to the scrabblings around the walls and rafters, a glint of cunning in his eye. After a moment or two he gestured to the woodsman for his walking stick, then advanced into a corner and peered up into the shadows. The rest of the party had by now stopped what they were doing and were watching him.

For some time he remained quite still, then with a sudden

grotesque contortion of his cheeks, he sucked in through clenched teeth to emit a piercingly high-pitched, reedy whistle. A second later the stick swept viciously over his head and out of the darkness plummeted a dead rat. There was an appreciative titter, at which the thatcher smirked and took a little bow.

He then moved to the opposite corner and shortly repeated the exercise, this time to loud applause, the loudest coming from the merchant's wife. She seemed to have recovered herself completely and began to egg him on with a most unmatronly ferocity. Her expression declared that she held rodents responsible for all the ills of the world.

This was scarcely pious or sober, I thought, as I left them noisily swatting rats and made my way through the courtyard and out into the street, but neither was the general air of the town, for that matter. It was more as if the place had been given over to a permanent fair. Even here, almost on the outskirts, there were people strolling, sitting in the sunshine, buying dubious relics, curios and sweetmeats from stalls or simple trestles placed across the doorways of dwellings. There was evidently no lack of enterprise amongst the inhabitants.

Was Ellen one of them? Would she be glad to see me if she were? Was the feeling that I had now been harbouring for so long a real feeling, or merely some wild imagining that had rooted itself in my conscience? All at once I felt acutely nervous. I raised my fingers to the stone I had been wearing round my neck since the old monk had given it to me, and startled myself by silently asking Roland to make sure that everything turned out well – whatever that meant . . .

'Carn stannit when 'e does 'at.'

I turned to see the wench standing beside me. She was scuffing the ground with one toe and scowling.

'Beg your pardon?'

'My pa. Carn stannit when 'e swipes 'em rats. Always doin' it, too. Makes me 'eave.'

'I suppose we'll sleep better.'

'Pilgrimages!' She spat. 'Piss on 'em, I say.'

'I wouldn't know.'

'What you 'ere for then, eh?'

'Well . . .'

'Go on, tell us. I can keep a secret!'

'It's no secret. I've just come to meet someone.'

'Now? Right now?'

'N-no. Not just yet.'

'Can I walk with you, then?'

I could not imagine anyone in the party whose company I desired less. It did, however, offer me an excuse to delay going in search of Ellen a little longer. I needed more time, to build myself up to it. I nodded, whereupon she took my arm and hauled me away down the street.

Before Saint Huctred's return to celebrity, we could no doubt have walked from one end of the town to the other in a few minutes. Now, there were so many pilgrims and so much coming and going of the produce and goods required to feed, water and otherwise relieve them of their coin, that in some places the tide of pedestrians, animals and conveyances had almost reached a standstill. Then there were the diversions, and my companion, I soon discovered, was easily diverted. Every few paces I found myself being dragged through the throng to inspect something that had caught her eye on the other side of the street: another collection of gewgaws, more brine-filled jars of the miraculous mushrooms, another pile of wood shavings masquerading, with full provenance, as pieces of the True Cross. In due course she tugged me to a halt at a street corner where a chained bear swayed somnolently from side to side to the beat of a drum, apparently oblivious to the antics of the small ape standing on its shaggy shoulders and fondling itself with a look of rapt concentration. Peg, for that was her name, found this hugely entertaining and, I guessed from the dewy look that crept over her, more than a little titillating.

It was a relief to be spared her prattle. My mind wandered off, endeavouring vaguely to insinuate itself into the fabric of the place in case there was some hint of Ellen's presence to be detected there. This was the sort of thing Roland would have been able to tell . . . I checked myself. It was all too easy to invest people with great accomplishments when they were no longer there to disappoint you. And what seemed most important was to retain a true memory of Roland, not let the passing of time inflate him, since the things I had loved about him, still loved about him, were the simple things, the things I hoped that I also in some measure possessed . . .

A hand fell squarely on my shoulder.

'The map-maker!'

It was the knight's nephew. He rose and fell on the balls of his feet, smiling broadly and full of restless energy.

'You look solemn, my friend . . .' His eyes strayed to the ape, now quivering with ecstasy on the bear's shoulders, and he grimaced. 'Dirty little brute! Saw quite enough of that on campaign. Nearest whore was a hundred leagues off . . . Something wrong?'

'No. Just remembering someone.'

'Ah.' He gave an appreciative nod and fell silent for a moment, then thrust out his hand: 'Humphrey, my name.'

'Cre-istopher.' On leaving the monastery for the second time, I had resolved to abandon Creb, but at certain moments when I was preoccupied, it still tended to slip out. We shook hands and he looked at me with a trace of amusement, as if this were some quaint provincial pronunciation he had not encountered before.

'Do you remember well, Christopher – have a good memory, that is?'

'For some things.'

'Directions . . . ?' He grinned engagingly.

I nodded.

'Hmm. They interest me, you know, these maps of yours. How about a pot of ale together? Then you can tell me more. What do you say?'

A further delay? My resolve still needed firming and ale might help.

'Yes, thank you, but . . .' I glanced at Peg.

He grinned again, but without malice. 'Oho! Latched onto you now, has she? She's all right, Peg. Just got her brain in the wrong place.' He leant across and whacked her backside.

'Oy!' She spun round furiously, then saw who it was and let slip a coy smile.

'Ale, Peg?'

Her eyes lit up and without a moment's hesitation she fell into step between us, the ape already forgotten.

In the centre of the town, an inn and an alehouse stood on opposite corners of the broad crossroads which now doubled as market-place and terminus for the steady streams of pilgrims who were either recently arrived or preparing to depart for the shrine.

In quieter times, perhaps, the two establishments might have vied for custom. Now, there was more than enough for both as people spilled noisily from the two doorways to mingle, ale in hand, on the sunlit crossroads. If we intended to talk, there was clearly no point venturing inside. Humphrey disappeared into the mêlée in search of drink, leaving Peg and I to find a seat. This we eventually did, squeezing ourselves onto the plinth of the stone cross at the centre of the crossroads.

I sat foward, apprehensively scanning the crowd as we waited for him to return. Nothing familiar caught my eye and as I continued to gaze into the noisy, shambling throng, I found myself beginning to wonder what precisely it was that had drawn these people here. A pile of bones, or some purported organ pickled in vinegar? Perhaps merely a slab of stone or some tawdry heap of masonry? And what would occupy their prayers? Relief from their bunions? Success in their amorous pursuits? The sudden and unexplained disappearance of their creditors? Devotion was conspicuously absent from the sea of faces before me, animated as they were by sunshine, ale and the companionship of the adventure, and it occurred to me, not for the first time, that the human capacity for belief was indeed a most unfathomable and tortuous affair . . .

Humphrey reappeared. At another time I might have sought his views on the subject – he was himself a pilgrim, after all – but now I felt the need to conserve energy and emotion. I leaned back, sipped my ale and let him quiz me about the maps.

He made intelligent observations about scale and contour, quickly perceived the problems of identifying the sometimes less obvious but all-important landmarks, and was politely critical of my system of dividing the page into squares which read left to right, two to a row, four rows to a page. It would be better, he said, if the route could be presented unbroken on the page, running from top to bottom. It would double the number of pages, of course, but it would be easier to follow.

I agreed with him in principle, but forebore from explaining that the same had already occurred to me, and furthermore that the tablets now gave me the means to sketch at least a week's travelling before committing it to parchment. As it was, I felt that the existing form allowed the maps to assume the character of a daily journal which, in an abstract but important way, had come to represent my own progress: the sharpening of my awareness, my increasing ability to answer my own questions, my growing understanding of what was taking place around me – all things whose ideal unit both of record and measurement, given the eventful nature of journeys, was one day at a time and no more. What is more, I had felt strangely superstitious about breaking the rhythm of my work in the middle of this particular journey.

Now he began to expound on the uses he could have found for such a thing in the foreign lands he had visited, and as I listened to him brimming with enthusiasm, I began to realise for the first time that there might be some wider value to what I had so far regarded as little more than an engaging, if somewhat eccentric, pastime; one which held certain obscure meaning for me, it was true, but very little, it seemed likely, for anyone else. I soon had a dozen sheets of parchment spread out on the ground at my feet (I kept my sack with me at all times and guarded it as if it contained the Holy Grail), and a small crowd had gathered to gape in the way of people with time on their hands.

They were proficient now, my maps, I was ready to admit. The lines were clean and straight; turnings, forks and intersections were clearly indicated; there was enough detail to pinpoint a position every few miles, but not so much as to distract or confuse, and I had refined the regular symbols (including the ones I had created for my own amusement – a foaming jug for an alehouse, a pair of sharp horns for a fierce bull, a dripping boot for a ford) to the point where I believed them more or less universally recognisable.

Humphrey picked one up and held it for a moment, then said: 'May I try something? An experiment?'

I hesitated. 'You won't harm it . . .?'

'Never!' He was already on his feet, smiling genially as he approached a dour-looking bystander.

'You, sir. Be so good as to answer me a question. Are you from these parts?'

The bystander shook his head.

'Did you travel far to get here?'

'Soom way, aye.'

'From the north?'

'Aye.'

'Fine country, fine country . . . and may I ask how you found your way here?'

The bystander frowned. 'By t'road.'

Humphrey nodded appreciatively and paused for a moment, then: 'Did you know the road?'

'Nope.'

'So . . . how were you able to find this place?'

The fellow's frown deepened, as if at some disagreeable recollection. 'Ah found it.'

'But with some difficulty?'

'Could say that.'

A sympathetic 'Mmm', another pause, then: 'I believe, good sir, that one of these might have assisted you.' As if it were the most succulent of sweetmeats, Humphrey delicately proffered the parchment. 'It's a map. Tells you how to get to where you're going from where you started. An uncommonly practical thing. Here, take it. See for yourself.' He pressed it gently yet firmly into the man's hand, then stood back, a respectful look on his face.

The bystander glanced down briefly, suspiciously, then declared: 'Don't read, m'self.'

'No need to. It's all drawings, pictures – so everyone can use it. Allow me to show you.'

Humphrey stood closer and began to trace through the squares with his finger, pausing to identify a village here, a bridge there, a note of reassurance in his voice now. In the fellow's eyes, suspicion yielded reluctantly to comprehension.

'Now. What say you? A fine thing for a traveller to have?'

'Mebbe.'

'And if your business took you often to different parts, would you pay good coin for one?'

'Depends where I was going . . .'

'Of course, of course. But in principle, a traveller's aid worth paying for – at a modest price. Would you not say?'

'Mebbe, aye.'

'And the better off you'd be, too!' said Humphrey, cheerfully disregarding the native caution. He smiled broadly and shook the fellow's hand. For a moment the bystander hovered uncertainly, wondering whether he was now going to be invited to buy one. When it became clear that he was not, he turned and went on his way, nodding to himself as if he had been somehow enriched by this brief but bewildering encounter. To my astonishment, the crowd applauded.

'Now,' said Humphrey, turning to another, 'what about you, pretty mistress . . .?'

For half an hour, he manipulated a swelling audience like potter's clay. It was a masterly display of charm and cajolery, teasing and sympathy, coaxing and soothing, all polished to sparkling effect by the ingenuous and good-natured manner in which it was delivered. As he warmed to the task, he seemed almost to gain physical stature, his energy spilling out into the space around him and wafting captivation amongst the crowd. He was story-teller, magician, master-pedlar, scholar, veteran traveller and soldier all in one and I, like all others present, was spellbound; doubly so since they were my maps around which his performance revolved. Even Peg was roused from wherever her reverie had taken her and stood up to watch with solemn interest.

When at length he spread his hands, flashed an apologetic but conclusive smile, and began replacing the pages carefully in my satchel, the crowd was standing a dozen deep around the foot of the cross. They would surely have remained there all day had he continued, but now they began to drift off in a buzz of conversation, empty cups and tankards dangling from their hands.

Humphrey turned to me, wiping sweat from his forehead. 'Well . . .?'

I could no longer conceal my exhilaration. 'Magnificent! A marvel . . . but . . .'

His eyebrows lifted. 'What but?'

'But . . . because of the maps . . .? Or just because you could sell shit to a swineherd . . .?'

We looked at one another and began to laugh.

'Both, by God!' He thumped me on the back. 'You have the parchment. I have the patter. What could be better? Peter Mapper . . . picked a peck . . . of pickled parchment!'

Tears were rolling down our cheeks. Peg was grinning in the way of someone who has not quite understood the joke.

'We could . . . we could . . . make . . . money . . . doing this,' Humphrey spluttered.

I nodded, my breath trapped somewhere in my gullet by the tide of hilarity and elation that was roaring through me. Wave after dizzying wave of opportunity was breaking in my mind, obliterating every other thought. I reached out and grabbed the startled Peg around the waist, heaving her off her feet, and she – with all the reflexes of the true opportunist – flung her arms obligingly around my neck and pressed her body firmly against me. Humphrey seized the moment to whack her rump again and she giggled and wriggled and squirmed. Suddenly, through it all, I had the unmistakable sense of being watched.

I turned as far as I could with Peg still clamped around my neck and for a moment found myself looking directly at Ellen. She was some distance off and had paused, in mid-step it appeared, as she made her way between two groups of bystanders. I do not know how long she had been looking at me, but the moment our eyes met she turned sharply away and walked on hurriedly, to disappear down an alley.

I dislodged Peg without ceremony and she landed on the ground with a startled grunt. I sank to the stone with a feeling of terrible emptiness, as if all the energy had suddenly been sucked from me. I tried to recall the look in Ellen's eye at the moment our gazes had met. But all I could see was the deliberate and abrupt closing of a shutter. I had not even registered what she was wearing.

Humphrey glanced at me quizzically, but said nothing. On Peg's face, however, bewilderment and dismay were now yielding to anger. Still sitting on the ground, she kicked viciously at my shin.

'That *hurt*, you shite! Whadd'you do that for? Eh? Whadd'you do that for? Like throwin' girls about, do you? Eh? Knob-licker! Arse-worm!'

I had no stomach for Peg's abuse. I stood up with a mumbled apology, nodded to Humphrey and set off on weak legs for the alley down which Ellen had disappeared.

Twenty

would have fared much better if I had made enquiries in the inn, but for the time being logic had abdicated to an engulfing sense of anxiety and gloom and I wandered around the town with little idea of where I was or where I was going, driven only by the vague expectation that instinct would lead me to her. It did not, of course. And by the time it occurred to me to ask someone, I had found my way onto the road which ran out in the direction of the shrine, and there, every person I stopped was a pilgrim.

I walked back towards the centre of the town, trying to distinguish inhabitants from visitors, but of the few I picked correctly, not one knew anything of a young seamstress, nor indeed of any young woman answering to Ellen's name and description. At length, however, I was told where I might find the doyenne of the local seamstresses. If anyone knew, I was assured, she would.

I reached her house without difficulty and knocked at the door. It was opened by a pale, skinny youth in the full eruption of adolescence, whose sullen manner and lack-lustre gaze suggested he spent too much time alone with thoughts as unwholesome as his appearance. He grunted that his mother was out and turned to step inside again, but before he could close the door in my face, I took the chance to explain why it was I needed to speak to her.

He paused and propped himself against the jamb. Something sly and lecherous slid into his eyes, and I regretted the question instantly. He knew of Ellen, he said, fingering a pustule on his neck. What did I want with her? I replied that that was no concern of his; I just wanted to know where she lived. He gave me a long, insolent look, then shrugged and obliged.

The address, surprisingly, was in the affluent quarter at the centre of town, where half-timbered houses rose to two or three storeys, crowding in over the cobblestoned streets to form deeply shadowed, echoing tunnels. But away from these dark, noisy rat-runs, through arches or behind tall wooden gates, were private courtyards and storehouses, mews and stables. It was on the threshold of one of these, a few minutes later, that I stopped. I looked up at the building before me, then at those on either side, secretly hoping there might be some ambiguity about the youth's directions. But the features he had described were unequivocal – an imposing pair of studded wooden gates and a narrow balcony at the first floor with the distinction of a climbing mouse carved on each alternate wooden baluster.

The gates were ajar and I peered tentatively into a pleasant, sunlit space, a paved yard with well tended flower-beds around the edges and a waterfall of newly laundered lace hanging from a line across the centre. In places, the lace reached almost to the ground like a huge white veil of the most delicate design, and as a sudden breeze ruffled it, causing the intricate shadows beneath to shift and swim, I might, at another time, have paused to savour such unexpected beauty. But now I was concerned only with the voices beyond.

I took a couple of steps forward, stopped and listened. Noises from the street drifted in. The conversation rose and fell but continued. I stepped closer. One of the voices was Ellen's. There was a brittle edge to it. I walked up to the lace and squinted through, unseen. Against the high wall opposite was a small outhouse, thatched and whitewashed like a cottage. In front of it, Ellen sat on a stool, talking. A youngish man stood listening, looking at her intently. Ellen smiled at him, a forced smile, from the mouth only. Then she dropped her gaze to the ground. The man squatted down beside her and put his hands on hers. He talked in a low voice. Then he felt in his pocket, pulled something out and closed her hands around it. Ellen let her head slip forward onto his shoulder. Another sound intruded, the muffled cry of an infant. Ellen looked at the man with a weary smile. She said something inaudible. Then she rose and went into the cottage. He also rose and left by a doorway into the main house. Some moments later Ellen came out, carrying a child. The child still cried and she fussed over it, cooing, kissing its head, stroking its cheeks. She sat down on the stool with the child in the crook of her arm. Then she lifted one hand. The long, slim fingers clasped two gold coins. She twisted them and they glinted in the sunlight. The child stopped crying.

My head was beginning to swim. My eyes were hot and stinging. I turned away and made for the gate. As I left, I could hear Ellen starting to sing softly. Snatches of the melody reached me. It was the song of the young shepherd and his girl.

Outside in the shadowy street, I removed the leather thong from around my neck and let it fall to the ground.

The pebble clicked as it hit the cobbles.

The revelation came to me as I sat staring into my cup in the alehouse at the crossroads. I had been there for an hour or so and was already quite drunk.

I must have arrived at some critical moment where alcohol and misery react to ignite strange concatenations in the brain, for it occurred to me suddenly that I had not so much chosen to come here as been guided, and that whatever hand was responsible had steered me not, after all, to Ellen, but to Humphrey. I was to throw in my lot with him and become a proper map-maker. I waited a while to see if the notion would slither away into the ale-ridden haze, but far from doing so, it seemed to gain plausibility with every second. I drained my cup and left the alehouse in the falling dusk, thinking fixedly of a future vibrant with journeys and the rewards to be had from them. By the time I returned to the inn, the misery had temporarily departed and I was suffused with an inebriated sense of wonder at my capacity for optimism.

There was no one in our sleeping quarters, but sounds of laughter from across the courtyard led me to the general vicinity of the kitchen and from there to a shabby room which appeared to have been commandeered by my new companions. It was unswept and unfurnished apart from a couple of benches, some stools and an elderly table in danger of collapsing under the weight of a brace of roast heron and a prodigious quantity of wine.

With their faces slack and shining in the lamplight, it was apparent that they had already been drinking for some time, despite the friar's earlier admonishments. It was a good four hours since I had left them swatting rats. The thatcher, by the look of him, had had his nose in the bottle all the while, the woodsman also. The others had all reached varying stages of conviviality; even the friar appeared flushed, his eyes endeavouring harder than ever to escape from one another.

At one end of the table, the merchant, the cordwainer and Humphrey were locked in loud and animated dispute which, I gathered, concerned the business of running an army. As I approached, the merchant beamed at me and mumbled something welcoming, while Humphrey made space between himself and the cordwainer, winked and passed me a platter. Opposite us, Agnes from the village patiently endured a monologue from the merchant's wife on the iniquitous ways of servants. As I sat down, Agnes smiled at me and draped an affectionate arm across the shoulder of her husband whose weatherbeaten brow glistened with sweat as he attended silently and methodically to the despatch of a drum-stick so large that it obscured half his face. At the far end of the table, the friar listened in evident bewilderment to some rambling tale from the thatcher who punctuated his narrative with frequent slappings of his thigh and brayings of crazed laughter. Between them, bored to distraction, sat Peg. She caught my eye and glared vengefully.

I ignored her and turned my attention to the mound of greyish, gamey meat on my platter. Unsurprisingly in an establishment such as this, the birds had been hung almost to the point of putrefaction. But no one else seemed to be troubled and I soon found that, washed down with plenty of wine, the meat was palatable enough. In any case, all these upsets had left me ravenous.

'. . . and that way,' concluded the merchant, belching loudly, 'you'd turn a devilish good profit, if I do say so m'self.' He ignored a disapproving glance from his wife and leant back on his stool with a venal smile.

'Humph!' said Humphrey. 'Trade, profits, interest. All very well for grain, I don't doubt, but mark my word,' he wagged his finger vaguely, 'you'd never make a commander – at last, you wouldn't least long as one. Ha, ha! Last, least . . . An' whathink you, map-maker? Our good merchant here – as marshall of the field?'

It was an improbable notion, I had to admit, but as I was seeking a tactful reply, the cordwainer gave a cheerful snort of derision: 'Far too fat!'

For a moment the merchant hovered on the brink of outrage

then, apparently, decided that he hadn't the energy for it. He patted his paunch affectionately and returned a rueful grin.

'And far too crafty,' added the cordwainer in a stage whisper.

'Speak for y'self, *craftsman*,' retorted the merchant, entering into the spirit of the exchange.

Humphrey pulled a face and turned towards me. 'How goes it, then?'

I nodded, my mouth full of heron, and grunted that all was well.

'Good! So whadd'you think? Maps? Money?'

'You tell me.'

He smiled conspiratorially. 'Well – let's say I've developed a taste for travel. Been thinking – since this afternoon. Pilgrims. Lost souls really, don't y'gree? Need a lot of guidance, pilgrims.' He took an energetic swig of wine. 'Holy Land – now there's a place for a pilgrimage. Trouble is, it's full of heathens. Scrofulous lot. Don't understand a word we say, and don't know a shit from a shave anyway. No point asking one of them how to find a shrine. Hump your baggage. Allah! Allah! Hump you too, if you don't watch out. But take you there? Huh! And that's only when you're there. Got to *get* there first – the Holy Land, I mean.' He paused and cocked an eyebrow.

'Go on,' I said.

'Look – I've had a bucketful this evening, Chroosipher – but don't mistake me – I'm serious – anyway, think I could twist his arm for the necessary . . .' He peered into his cup, nodding to himself.

'Whose arm?'

'Whose arm? Hmm . . . Uncle's! Uncle, of course. The good knight. M'guardian of erst. Son he never had, see. Makes you weep, doesn't it! Proud old pikestaff though – proud of me, I mean. Fond of *him*, too. Always kep' my nose clean with him . . . Hey, Peg! Going already? Gissakiss. Go on! Ahhh, that's it! Sweet dreams, pretty . . . Just wants a bit of attention, our Peg. Who'd blame her with a father like that . . .'

I looked up to see the thatcher dead to the world, his head sideways on the table and his mouth wide open.

'What was I saying?'

'Your uncle – and the necessary . . .'

'Uncle . . . Mmm, well, salted it away over the years, know he has. Did his crusading, of course – properly you know – all chivalry and pen – er – penniless, like the rest of them. But he was away a lot – most of the time – and it wasn't all crusading. Wars . . .' He waved his hand vaguely. 'Someone's got to fight 'em, haven't they? If I'd picked the right one, I'd have the wherral – whatsit – m'self. But I didn't. Pity. Anyway, he'll like it – Uncle. Know he will. Probably want to come with us. Too old though, thank Christ.'

'Hmm. How much then? What would we need?'

'Oh . . . if we took it in stages . . . few months at a time? Let's think . . .'

And so it went on.

One by one, the others retired to bed or, following the thatcher's example fell asleep at the table. Humphrey and I drank steadily and our conversation became less and less coherent. At some point in the evening we finally exhausted all possible consideration of our future and moved onto other things, of which I remember only a garbled tale involving a dying companion-at-arms, a trio of foreign whores, a badger and a dose of the pox – all of which conspired in some obscure way to explain his presence here.

By the time we came to snuff the remaining lamps and heave ourselves up from the table, I felt as if I had gone without sleep for several days. My mouth was dry, my vision blurred and my head pounded. Worse than that though, the drink had sluiced away all my earlier resolve; images of Ellen danced about unsteadily in my mind's eye, and I felt at a very low ebb indeed.

We were halfway across the courtyard when I remembered my sack, which I had placed under the bench on my arrival. I made my way back to the dining room, stumbling into walls and doorways in the darkness, and spent several minutes on my hands and knees under the table, mistakenly grasping at feet, before finally having to admit that it was not there. At that moment I wanted nothing so much as to retire to bed, but the self-imposed conditioning of the last few months refused to allow it. I stood in the darkened room, listening to the snores and trying hopelessly to think what could have become of it, then set off again to enlist Humphrey's help.

It was a warm, overcast night. Despite the holes in the roof, the interior of the outbuilding was almost as inky dark as the other room. Humphrey must have fallen asleep instantly for no one stirred and it was some while before my eyes adjusted sufficiently to pick him out amongst the shadowy, recumbent forms. As I began to tread gingerly across the straw towards him, something made me pause to count the other sleepers. There were four apart from him. Three were still sprawled at the table in the other room – the thatcher, the merchant and the woodsman. So someone was missing. I attempted to identify the other dark shapes. A wheezing tumulus in the far corner – the merchant's wife. A long, spare form, arm thrown across its face and one sandalled foot protruding from beneath the blanket – the friar. A smaller shape lying neatly on one side with the bejewelled fingers of one hand clasping the blanket to its throat – the cordwainer. And a hedgehog-like ball with its back to the centre of the room and its head completely covered. I tiptoed across, held my breath, reached out and pulled back the blanket an inch.

Agnes.

So it *was* Peg, as I had guessed.

God damn her! She could so easily have taken it during her brief exchange with Humphrey . . .

This was the last straw. I went quietly outside and sat down against the wall. God damn her again! I was already feeling dizzy and a little queasy, and now my eyes had begun to prickle with self-pitying tears. Thinking made my head spin. But I had to think. What would she do with it? What would I do with it if I were her, if I wanted revenge for a slight to my dignity? Destroy it? Simply throw it away somewhere? God alone knew. I couldn't read the bitch's mind. Certainly not in my present state. The only thing to do was to go after her, spend another few hours wandering around this Christ-forsaken town looking for another woman. Then, by the huc of Saint Cocktred, thump her. Thump her whether she'd still got it or not.

I sat there for another few minutes, working up a wholesome rage, then rose and stumbled out into the darkened street.

In recent months I had come to know towns, to understand how they worked and to appreciate them. I had learnt to enjoy the hustle and bustle, the garish sights and incessant noise, the invisible but pervasive pulse of commerce and all the other undercurrents of ebb and flow when many people are thrown together. By day, I found towns exhilarating. But come the evening, it seemed that some strange, almost sinister transformation began to occur, a dissolution of all the sparkling veneer, revealing the truly base nature of urbanity. At night, towns echoed the dark, the tortuous and the sordid facets of life, as if all the human disquiet and desolation which was held in check during the daylight hours found release after sunset and spilled out from its sleeping hosts to stalk the empty streets and shroud the darkened buildings. And in some strange way, the presence here behind closed shutters and doors of so many pilgrims with all their pent yearnings and fears, their curious idolatries and superstitions, made this town seem even eerier, more squalid than most.

I set out down the middle of the street so as to keep my distance from the shadowy recesses lining the way, and tripped into a pothole. A large dog, all ribs and teeth, lifted its head from a pile of offal and growled as I approached. My detour took me beneath some houses where an upper storey window creaked open and a rain of night-soil pattered to the ground behind me, splashing my heels. I stepped over a beggar sleeping in a doorway and trod in something vile.

There was only the occasional burst of muted laughter, together with the faint whiff of woodsmoke, surprising on a warm night such as this, to remind me there were still people about and that I was not running the gauntlet of the deserted streets purely to nourish my ill-humour. Somewhere out there was Peg, Peg of the small brain and the big cleavage – into which I was going to jam her grubby nose when I found her.

I was drawing close to the centre now. A couple were locked in deep embrace in the shadows of a doorway. A solitary figure swayed towards me and grunted as it passed. Lights wavered up ahead, from the direction of the crossroads. Voices carried down the street. So perhaps there was still life at the inn or at the alehouse. And if there was, that was where Peg would be. Trying to sell herself, more than likely . . . no, not herself . . . the maps . . . or maybe herself and the maps! Peg who had watched Humphrey's performance with such uncharacteristic attention. Peg of the not-quite-so-small brain.

Roland would have been proud of me.

I hurried on, the clatter of my quickened footsteps echoed by a regular jabbing within my skull, but I ignored it, ignored the sleeping bodies sprawled at the base of the stone cross, the ground around it strewn with the detritus of the evening's revelry, the crusty puddle of drying vomit on the doorstep, and entered the inn.

In a candlelit alcove facing the door, an aproned man sat with his legs astride a small barrel, counting a glinting mound of coins. He glanced up at me and shook his head peremptorily.

'Just looking for someone. Anyone still here?'

He flicked his head to the right and continued counting.

The room he indicated was pitch dark, the air stale with the odours of drink and perspiration and the faint musk of furtive carnality. A lamp, guttering low in its bracket in the passage, cast enough light through the door to reveal a handful of empty stools and a couple of small tables, upon one of which lay an abandoned set of dice. But although I could see no further, I sensed that the room was not small. Nor was it empty. Its occupants, wherever they were, were sleeping noisily. A candle stub lay on the floor. I lit it at the lamp in the passage, then made my way cautiously forward, hot wax dripping onto my fingers. The nearest sounds came from beneath a bench. The candle scarcely lit my feet and only by crouching was I able to discern the form of an elderly man lying with his nose pressed up against the wainscot, rhythmically grinding his teeth. I moved on like a thief, stepping unsteadily around the furniture in my tiny pool of light. A second comatose customer appeared to have slid to the foot of the wall with his head on his knees. A third was stretched out on top of a bench. And then I caught a soft giggle from the furthest corner.

I tiptoed forward, cupping the candle with my hand, then stopped and held it up. At the very limit of its radiance, half in shadow, Peg was straining backwards on an upturned keg, her bodice undone and her breasts bursting forth while between her spread thighs her skirts appeared to be undulating in a curious fashion. Her eyes were screwed shut and for a moment I hesitated, ashamed and resentful at finding myself cast in the role of voyeur once again. I reminded myself of what I had come for, stepped forward and grabbed her arm. She opened her eyes wide and screeched like a jay. I shook her. 'My sack! I want . . .' Her free arm came round like a whip and fingernails raked my cheek. I slapped her. She screeched again. There was a crash of furniture in the darkness behind us. A figure was emerging from beneath her skirts. She was certainly not being paid for *that*, some distant voice was telling me. I dropped the candle and pinioned Peg by the arms. I was shaking her now and yelling. The shadows around us began to dance at the glimmering approach of a lamp. The figure had backed under a table and seemed to be stuck there. Purposeful footsteps accompanied the lamp. Peg writhed and continued to screech. The figure heaved the table aside, straightened itself and wiped its mouth with the back of its hand. It was immense and bald and its voice came from its boots: 'You . . . little . . . turd!'

I let go of Peg and began to edge away. I saw the figure raise a shadowy fist, big as a loaf. The footsteps were right behind me. I backed into something soft but unyielding. The fist came towards me in a blur of speed and I thought I heard my skull crack . . .

Twenty-one

hen uttered in a particular way, certain words have the power to penetrate the most unreceptive ears and strike at the very soul. Fire is one of them.

Despite the agony in my head, despite the bile rising in my gorge, despite the darkness which enveloped me, I sensed raw panic sweeping like a gale around the inn. The sounds of commotion, of people running, shouting, of wheels clattering, animals whinnying, bellowing, grunting – all were sharpened on the whetstone of fear.

I scrambled up from amongst the wreckage of a table and a couple of stools. My head and stomach reeled in concert and I was violently sick. When I was able to move again, I felt my way out of the darkened room and into the street where I was instantly assailed by more than my senses could bear. The chaos made me dizzy. Fearing that I might vomit again, I retreated into the doorway of the inn and stood still.

Overhead, the sky was stained as if by the arrival of some fiery, apocalyptic dawn, spreading unnaturally from the west. There, at its source, a deep glow threw the rooftops into flickering relief and hurled sparks high into the night like sprays of incandescent chaff. The air was heavy, almost sickly, with the woodsmoke I had smelt earlier, the sharp odour of burning thatch, the tang of roasting spices and a host of other unknown pungencies released by the blaze. Before me, the crossroads had degenerated into a scene of utter confusion. At the foot of the cross, eerily shadowed by the ruddy light, pilgrims milled in gesticulating groups, their faces puffy with sleep, frightened incomprehension in their eyes, whilst around them streams of dishevelled townspeople flowed chaotically hither and thither. Some, carrying buckets, pots, whatever receptacles they had picked up in their haste, headed for the conflagration; others, burdened down with still-sleeping children, animals, household possessions, fled from it. An old man in a nightshirt was attempting to drive a flock of geese through the throng. A donkey pulling an empty cart, to which were attached a goat and two sheep, had become separated from its owner and stood blocking the entrance to a side street. Its forlorn braying pierced the ambient cacophony of shouts and curses, children wailing, doors slamming, mothers shrieking for missing infants, feet clattering and the collective groans and cries of the crowd as the ominously muted crackle and dull roar of flames swelled and retreated.

I felt dazed and weak and curiously detached from the frantic activity around me. I could observe with clarity but found myself reluctant to participate in the destiny of this town that had caused me so much grief. The danger was real enough, and so, therefore, was the fear that gripped the inhabitants – fire is the extreme peril for any village or town built predominantly of timber, as most are. But for the time being I was safe enough where I was and lacked the spirit to do anything except remain propped in the doorway, watching the drama unfold.

Someone burst from the inn behind me and nearly knocked me over. The innkeeper. He swore and bent to retrieve one of the two pails he had been carrying, then recognised me and glared.

'You!' He thrust the pail into my hand. 'There, you can make amends with that. Get going!'

I found myself following him meekly into the crowd.

We left the chaos of the crossroads and joined the urgent procession of fire-fighters, hurrying with their buckets and picks and axes through the narrow, weirdly shadowed streets towards the blaze. Sudden gusts of heated air tugged at our clothes and blew tendrils of acrid smoke towards us, making our eyes smart. The sounds grew louder, more threatening, crackle and roar interspersed now with the rumble and thud of collapsing buildings, the sharp reports of disintegrating timbers, the boom of obscure explosions. The light flared and wavered and flickered crazily, making the houses above us seem to teeter against the bloody darkness – and suddenly I registered my surroundings, recognised the street we were in, realised where it was that the blaze had taken hold. I paused, steadying myself against the wall as some hidden reserve of energy unlocked itself and surged through me, driving away the pain in my head, the nausea, the feebleness of limb. I dropped the pail and ran like a madman in the direction of Ellen's house.

It was not far, but I took a chance nonetheless and doubled down an alleyway which disgorged me, a breathless minute or so later, into the street at right angles to the one in which Ellen lived. Thirty paces ahead of me lay the intersection, with a left turn for the part of the street where Ellen's house stood. Spreading up the same street from the right, the fire had almost reached the junction and, judging from the movement of the flames above the rooftops, was beginning to break back into the houses in front of me. On the nearest right-hand corner some of the fire-fighters were doing their best to drench the interior of a narrow dwelling with water; others, leaning out from the upper windows or straining up from within the roof, were endeavouring to dislodge the thatch with scythes and billhooks. On the opposite corner, another group were making furious progress with the demolition of a single-storey building like some sort of craftsman's workshop.

It was a scene that could have been transposed from hell – at once dreamlike, fantastic, yet stark with the potency of real danger. Overhead, the flames leapt twenty, thirty feet into the air, bursting through the clouds of thick smoke which billowed out-wards and upwards into the darkness. The heat, even at this distance, was intense, the noise thunderous and fearful. Stripped to the waist and glistening with sweat, the bucketeers had organised themselves into a moving chain which emerged from the courtyard of a nearby house, stretched down the street and disappeared beneath the continuous shower of loosened thatch into the building on the corner. The man at the front of the line, having discharged the contents of his bucket, fled back to the well, replenished it, passed it on up the line and took his place at the rear again. The activity was constant and frantic. It was mirrored in the cobbles, slicked with spilt water, and in the larger puddles which reflected the flameshot darkness above. The whole road seemed to shiver and dance in a state of livid, aqueous fragmenta-tion. Beyond the bucketeers, and more vividly lit by the approach-ing flames, the demolition gang laboured with maniacal urgency, tearing at the roof of the workshop, wrenching and hacking at the supporting timbers, feverishly hauling away the debris to deprive the fire of its fuel. Half visible through the smoke, a further group were frantically prising up the cobbles in an attempt, I imagined, to reach some underground watercourse.

One of the thatch-removers lost his grip of the window-frame and toppled backwards into the street, his head striking the ground with sickening force. Instantly, two members of the bucket-chain broke ranks, seized him under the arms and began dragging him to safety, just as a small dog hurtled round the corner with the continuous high-pitched yelping of a creature in great pain or terror. It skidded on the wet cobbles, became briefly entangled in the legs of the rescue party, then broke free and bolted on up the street towards me with its ears flat, tail between its legs and hindquarters low to the ground. I saw that one of its flanks was blackened and steaming, mottled with exposed patches of raw, pale flesh and a cloying smell momentarily filled my nostrils.

A sudden shout went up. For an instant, all activity halted. Then, as one, the demolition gang threw down their tools and fled into the shadows. The bucketeers and thatch-removers began to pour in panic from the house on the corner, racing up the street towards me. Something was rumbling, very close and very loud. The air above us quivered, then flared with a dreadful brightness. A blast of heat swept over us as the rumble erupted into a roar and a monstrous gout of flame burst from beneath the eaves of the building next to the house on the corner, surged over it, consuming what remained of the thatch as it went, leapt across the street and took hold of the roof of the next house in line, wreathing it with fire. Seconds later, all light was extinguished by the boiling clouds of dense black smoke that followed in the wake of the flames.

There was something in the smoke which made the lungs contract, some kind of resinous vapour that seared the eyes and seemed to strip the linings from nose and throat. Choking and coughing and half-blinded, I groped for the nearest wall and stumbled forward with only the heat at my back to tell me which way I was going. Around me, other people were calling out to each other for reassurance as they too fumbled in panic through the darkness.

I ran into someone ahead of me.

'Tom? Tom? The – uh, uh – pitch store, wonnit?'

'I'm not Tom.'

'Oh. Reckon it was, though – uh, uh – nothin' else'd've gone up like that. Fire must've broken back into Cooper's Yard. Uhh.'

A draught tugged at the smoke and it thinned a little. Through streaming eyes, I could just glimpse the end of the street. I waited to hear no more, pushing past him and haring back the way I had come.

I found the opposite end of Ellen's street, almost weeping with the ache in my lungs. I could see down to the crossroads I had just left. It was now a raging and almost unbroken wall of flame from which the fire was making its way towards me with a horrible inexorability, spreading up both sides of the street at once. It had not yet reached Ellen's house but it would within a matter of minutes. I stumbled the last few yards, gasping at the waves of heat pulsing towards me, and pushed through the doors into the courtyard.

The uppermost parts of the surrounding buildings palpitated with flaming shadows, but down at ground level it was dark and the air was still and almost cool. I closed the doors behind me. The din of the fire abated a little and I rested for a moment. The lace still hung prettily from its line, a phantom-like presence in the gloom. Beyond it, the little cottage was in darkness and so was the main house. All seemed quiet and peaceful. But it would not be so for very much longer.

I ran to the cottage and hammered on the door. There was no reply, so I went in. It was a single room with a bed and a small chest, a chair and a stool, a table with some cooking things on it, and a fireplace. There were clothes over the chair, a skirt and a shawl, and my heart turned when I saw them. But the room was empty.

For a bleak moment I remained there. Then I hurried back across the yard and stepped outside to be assailed once more by a maelstrom of light and noise, smoke and searing heat. In the short time I had been in the courtyard, the fire had reached the roof of the house opposite, sending smoke billowing upwards and a shower of embers raining down into the street. A dull glow became visible at the upper storey and a second later there was a sharp crack of exploding glass, followed by a roar as a tongue of flame burst with shocking ferocity through a window and leapt the intervening space to the roof of Ellen's house.

I took to my heels.

When I was twenty paces from the end of the street, someone came pelting round the corner towards me. In the flickering, smoky light, it was not until the figure was almost upon me that I realised who it was. She had the child tucked under one arm. Her clothes were dishevelled. She looked exhausted and utterly desperate. But she was still beautiful.

She stopped in front of me, shaking and gasping, her eyes wide with surprise.

'Creb? Oh . . . oh, Creb! Thank God!'

An expression of overwhelming relief fleeted across her face and it was all I could do not to reach out for her there and then, but almost as soon as it had come, the softness left her again and she asked fiercely: 'Our house. Has it reached it? The fire?'

'Just. A minute ago.'

'Oh, sweet Jesus! You must help me, Creb!' She took my hand and began to run. 'The child! Quick!'

'But . . . you have your child . . .'

I could barely keep pace with her.

'No, no! The other one. The child of the house. They're away. And the old nurse is half crazy. Stone deaf, too.'

'But – how do you know I know?' I was having to shout now. 'Where the house is, I mean?'

'The necklace. I found it in the street. Oh, run, Creb! For the love of God, run!'

I reached the house a little ahead of her and dodged beneath the flaming debris to burst through the doors into the courtyard. It was now as plangent and savagely lit as the street beyond, for the roof of the main house was burning furiously, as was that of the gallery on the upper part of the street frontage, above the doors. Ellen followed me in and although she did her best to control it, I could see now that she was shaking not merely with exhaustion, but with fear – and I remembered, all at once, the manner in which her own family had perished. I too was beginning to feel afraid, because I now had an inkling of what she wanted me to do.

The house had three storeys, the uppermost of which was already beginning to blaze as the fire gnawed its way downwards from the roof. She pointed to a window in the second storey, at the far end.

'There's a passage that runs all the way along. They'll be in the room across from that one.' She looked at me beseechingly, then attempted a smile and raised her fingers to my cheek.

'Christopher . . .'

If I was to do anything it had to be now, before what little nerve I had failed me. I crossed the courtyard with a watery feeling in my entrails and entered the house. The open side-door admitted enough wavering flamelight for me to see that I was in a flagstoned passage with a narrow wooden staircase leading upwards, almost directly ahead of me. The staircase, presumably, connected with the passage above, so I had merely to turn right at the top, go to the end and enter the left-hand room, retrieve nurse and child and come down again. A simple enough exercise, it seemed.

But an unlit, unfamiliar building with its top floor ablaze is a place pregnant with the possibility of grim surprise. And the first came when I was no more than three steps up the staircase, as the door slammed shut from some wayward draught, plunging me into darkness. The sudden absence of light seemed to accentuate the sounds coming from above me, as if something monstrous was feeding on the upper floor, sucking and chawing and snarling. Now my eyes began to prick and I smelt smoke.

I groped my way up the remaining steps and as I stumbled into the passage began calling loudly for the nurse, but soon succumbed to a fit of coughing as smoke scoured my windpipe. The noise was getting louder and things began to creak and crack in the rafters over my head. The air was stifling and my shirt became damp with sweat. I felt my way along the passage wall, calling out when I could, but even if there had been a reply, I could not have heard it. The texture of the wall changed. A door. I pushed and it opened a few inches then met some unyielding obstruction. There was the merest shift in the darkness around me as a meagre shaft of light stole into the passage, to be absorbed within a foot or two by swirls of smoke. It comforted me all the same and I continued, bent low to keep my head out of the worst of the vapour. I winced at a rending crash above me as more light appeared overhead. I could dimly see flames through a chink in the timbers, then smoke came billowing through, forcing me to drop to my knees for fear of asphyxiation. I crawled forward and at last my head met the wall at the end of the passage. I felt for the door to the left, opened it, scrambled inside and closed it again.

There was smoke here too, curling down from the rafters and

creeping in under the door. But the window was open, allowing the air to clear and the blaze of the adjacent building to illuminate the room with an uncompromising starkness. I dabbed the tears from my eyes and was able to see that the simple bed and the more elaborate crib at its foot were both empty. But a curious sound came from a shadowed corner beyond the bed, where the light could not reach. There, huddled on the floor in nightdress and cap, the old nurse was rocking the infant in her arms and moaning to herself as tears streamed down her cheeks. She glanced up at me and for an instant I recognised in her look the same fearful resignation I had seen on the face of Roland's aunt as she cowered in the bedchamber at the manor. My heart sank.

I moved across to her and tried to ease her up. At first I could not obtain a solid grip since she had the infant clasped to her bosom and both arms tight to her sides; when I did gain purchase, she went first quite limp, then rigid as a board. The sounds of burning now seemed to be coming from directly overhead and I began to glimpse ominous flickers of light under the door. But nothing I said, cursing or cajoling, made any difference. Without her co-operation I was unable to move her. I was on the point of tearing the infant from her and leaving her to the fate for which she was so clearly prepared, when, without warning, a cataclysm erupted overhead. A flaming roof-timber burst like the finger of God through the rafters, plunged down and smashed into the bed, cracking the oaken pallet as if it were made of twigs. Flames immediately started to lick and dance on the covers as smouldering debris followed through the gaping, smoking hole above and a patch of night sky became visible.

The old woman shrieked, hauled herself up and started to shuffle towards the door. The child, which had been whimpering softly, began to wail. I darted ahead and opened the door a crack. The passage was thick with a noxious smoke which set me coughing at once, but not so thick that I could not see flames flickering and curling through the beams, and a dull but steady glow at the far end. I found myself thinking of windows and knotted bedlinen, but I knew the old woman would be unable even to climb to the sill. If it were only the infant . . . but it was both of them and, terrified as I was, something would not let me abandon her.

I grasped her by the shoulder and propelled her out into the passage, praying that the glow remained beyond the staircase.

Immediately my breathing became stifled, my head began to swim. I blinked furiously but my eyes now watered and stung so much that I was virtually blind. I dropped to all fours, forcing the old woman down in front of me. To my astonishment she contrived to clasp the child with only one arm and keep moving, making strange little mewing noises as she lumbered along the floor like some ancient, frightened, three-legged animal.

It was not far to the staircase, but the noise and heat and the flickering, choking smoke-filled darkness disorientated the senses to such an extent that, when I was not racked by spasms of coughing and waves of giddiness, some vestige of reason kept enquiring whether we were making any progress at all. Were it not for the hardness of the boards at knee and elbow, I would no longer have known. But somehow – God bless her! – the old woman continued to inch forward, feeling her way along the wall, and I followed behind her.

I sensed a little swirl in the smoke, felt the touch of a draught on my face, and opened my eyes for a second. We were level with the door I had earlier attempted to open. Ten paces to go. I roared encouragement to the old woman and she half turned, coughing and sobbing with fright, eyes tight shut – and suddenly the glow at the end of the passage flared. The air ahead of us ignited in a lung-searing flash, accompanied by a roar that seemed to fracture my ear-drums. It was some deep reflex, not selflessness, which hurled me on top of her, making me cover my head with my hands and hold my breath as, for an incalculable time, noise ransacked my skull and light drilled through closed eyelids, eyeballs, reaching right to the back of my brain. Then the boards beneath us shuddered with a thunderous impact. The heat was suddenly so intense that for a moment I imagined I could feel my skin blistering. A powerful smell of singed hair filled my nostrils. The old woman squirmed and cried out beneath me and I lifted my head to see that the whole far end of the passage was consumed with flame – not merely the passage, indeed, which no longer had a ceiling, but all that was left of the top of the house. And mere paces before us, across the mouth of the staircase, lay an impenetrable thicket of splintered rafters, loose thatch and shattered roof timbers, all blazing furiously.

I struggled upwards and backwards, hauling the old woman with me, aware that bedlinen was now our only possible recourse. As we turned from the flames I glanced at the doorway and for a mere fraction of a moment glimpsed what might have been the vaguest human outline, hovering amongst the eddies and swirls of clearing smoke. For an equally brief second, the hairs bristled at the nape of my neck and I felt a strange sensation of cool, clear, stillness in which lingered some presence as vague as the form in the smoke, yet as familiar as the feel of my own skin. But who or what it was, I could not say. Nor could I be certain that, shocked and terrified, I was not simply suffering a fugue of the wits.

But there was no time to ponder such things, for now the blaze had leapt forward from the debris at the top of the stairs, fanned by the new and plentiful supply of air, and the old woman was tugging in panic at my sleeve. On a sudden impulse I put my shoulder to the door and heaved; whatever had obstructed it so firmly at my first attempt appeared no longer to be there and it swung open without the least resistance, leaving me tottering on the threshold. I recovered myself, pulled the old woman into the room with me and slammed the door shut behind us, praying that what was left of the roof would hold up a little longer. The old nurse slumped to the floor, heaving and moaning and cradling the now hysterical child in her arms. From the look in her eyes, I feared she had exhausted her courage out there in the smoke and that this time I would not get her up again, especially when she understood what it was I planned to do. But I had to do it, nonetheless. I glanced around. There was no bed, nor any bedlinen, but the room seemed to be full of lace – shelves and cupboards and waist-high piles of it. We had stumbled into a storeroom.

With the fire roaring in the passage outside and flames beginning to lick under the door, I grasped a handful of lace from the top of the nearest pile and tugged at it. It was too flimsy. I tried another pile. Stronger, but of inadequate length. I needed more light. I reached for the window and was about to throw open the shutters when a deep vibration ran through the boards beneath my feet. For a dreadful instant the whole house seemed to shudder. The shutters creaked, then swung open of their own accord, and as the room was flooded with bloody, flickering light, it seemed as if the sky itself had caught fire. But it allowed me to see, against the wall, a pile of what appeared to be large rolls of delicate material, a few inches wide – the hems or borders for skirts perhaps. Surely there were enough there to contrive some kind of rope, however slight a single strand might be. To reach them I had to push a large, square bundle out of the way and as I heaved at it my foot struck something on the floor. I glanced down and my heart leapt as I saw a ring set into the boards. I pushed the bundle again, then hauled at the ring. A section of floor swung up on oiled hinges.

Whatever lay below was still in total darkness, but this was not the moment for caution. I shouted to the old woman to move up to the edge of the trap, that I was going to fetch something for her to climb down on. Without waiting for her response, I lowered myself into the hole and once I was fully extended, let myself drop. For a heart-stopping moment I fell freely through the void, then landed on something which rolled away with a clang and a clatter, pitching me heavily onto the floor. I scrambled up, bruised and shaken, and started to grope for the wall. But before I could find it my eyes began to acclimatise to the unsteady light admitted by the open trapdoor and I was able to see that I was in some kind of laundry room. There was more lace everywhere, along with the dim shapes of huge tubs and pans and kettles, mangles and drying racks, and there in the corner – praise God! – a ladder. I started to heave it into the centre of the room and there was a crash overhead, the light flared up. I wrestled the ladder to the rim of the trap and shinned up it as fast I could. As my head emerged, a wave of heat made me gasp. The door had fallen and the fire was raging in from the passage, spreading up the walls and starting to consume the lace. The old woman had edged back to the window and had one arm around the child and the other across her face. She had begun to tremble so violently that I feared she might be having a seizure. I shouted at her and mercifully she looked up.

'Give me the child! The child first – then I'll come back for you.' The heat was scorching the back of my neck.

She glanced at the approaching flames and shook her head.

I screamed at her: 'You'll die! Both of you!'

I started to climb out of the trap but was paralysed by the prolonged shudder which coursed through the whole building and set the timbers groaning around us.

'The roof's going! For the love of Christ, GIVE ME THE CHILD!'

She closed her eyes and her mouth worked in soundless prayer, then she rocked foward onto the knees and crawled towards me. I scrambled down the ladder, placed the infant without ceremony in a large copper pan, and climbed to the storeroom again. It was now alive with flames and the old woman gave a thin, high-pitched shriek as she teetered on the edge of the trap, flailing at the edge of her nightdress which had caught light. I held out my arm, caring little how I got her down just so long as we could both be free of this death-trap, but she turned around to step backwards onto the ladder, lost her footing and toppled onto me, throwing me against the edge of the trap and making the ladder twist alarmingly on one leg. For a giddy moment I thought I might steady it by bracing my feet against the rungs, then it twisted again and we both fell. This time I was not so lucky. I lay still for some moments, winded and sickened with pain. Then it came to me that the old woman was silent. I climbed unsteadily to my feet and in the flickering light saw at once that she had landed on her back across the rim of a huge washtub; from the way she was now draped, with her body awkwardly bent and her head lolling to one side, it was clear that somewhere – in her neck, or perhaps her back, it mattered little which – the brittle old bones had yielded fatally to the impact of her fall. I stood and stared at her in a daze, thinking only of how her nightdress was now no longer alight, when the building was once again riven by a long drawn-out shudder, like some huge animal in its death throes. A series of short, sharp reports overhead were followed by a dull but gathering rumble.

I came immediately to my senses, retrieved the baby from the pan where it now lay wide-eyed and silent, its gaze fixed in fascination on the flame-wreathed aperture of the trap, and left the laundry room. The passage beyond was brilliant with light as the fire stridently devoured the staircase, at the top of which, gradually dislodged by successive tremors, the blazing thicket of beams and rafters now teetered perilously. I fled for the outside door, expecting the thicket at any second to break free of whatever restrained it and come bounding and tumbling down the stairs. But instead of the leap of scorching flames at my face, I was greeted by a rush of cool air and kept running, clutching the infant to my breast, until I reached the far wall of the courtyard where I could run no further and slid in exhaustion to the ground. I became vaguely aware of Ellen kneeling at my side. I opened my eyes and as I did so the rumble grew to a roar and the whole building began slowly and deafeningly to collapse in a spew of smoke and flame, a welter of falling timbers and toppling masonry. Within moments it had been reduced to a flickering, smouldering ruin, less than one storey high.

The dawn came up like a rust-stain through the pall of smoke hanging over the town. It found us crowded, fifteen or twenty, into the downstairs room of a small house on the very outskirts. Everyone present bore some insignia of the night's endurance, and some more bravely than others. But I, for one, was happy. And so, I sensed, was Ellen. She sat on the floor opposite me, her head lolling in exhaustion over the swaddled form of the child of the house as it nestled to the soft swell of her breast, quietly taking her milk. Between us, an old man was stretched out on the bare boards, mouth agape, snoring gently. And across my lap, its head cradled in the crook of my arm, lay Ellen's child, quite calm and wide awake. It stared up at me with a look so profound and solemn that I found myself almost believing it understood who I was and what I was doing there.

Once in a while, Ellen's head gave a little jerk. She looked across at me through half-closed eyes and her smile sent a flood of warmth through my aching, weary body. I felt as if she could heal every sorrow, every pain I had ever known.

Twenty-two

I never did find Peg or my sack. Nor did I ever see Humphrey again. But I did become a proper map-maker, in so far as to gain part of a living from something renders it proper. And I did make love to Ellen; not once, but many, many times, the last of which was but a few hours ago. The sweetness of it still leaves me feeling that my spirit has been bathed in honey.

What was it that Roland said? 'Crebanellen . . . stars in heaven . . .' He knew, of course, in his uncanny way. And he knew also the manner in which we would continue. 'Far apart . . . but close at heart.' For my journeys take me away for most of the summer months, and after a while we both start to count the days until my homecoming, feel the gradual swell of expectation and longing; the little chills of fear and doubt also. We have not yet grown so accustomed to one another that we are like those pairs of old mittens which slip comfortably onto the hands at the beginning of winter, holes and snaggle-threads and all, as if they have never been absent. We need to re-discover one another each time and so far have always succeeded.

But it suits us too, I believe, this freedom to cast our separate beams of enquiring light upon the world around us. Ellen has her absorption with the child, a small but complete universe to which, however much I love them both, I accept that I can never be fully admitted. She also has her sewing which she has taken up again and is gaining her something of a reputation in the vicinity. There is now quite a steady stream of customers to the little stone, shingle-roofed cottage I have built up here on the hillside; then she makes a visit once a month or so to the small town nearby.

This grants her an independence of spirit and certain independence of means also. She takes great pride in her craft, and I am proud of her for it. I feel she is contented.

I am also deeply contented as now, in winter, when I work on my maps and write down my stories, read and learn, mend the house and, in exchange for produce or even a little money, do whatever stonework or joinery will please our neighbours. But when the spring comes, like a sailor who longs for the sea, I begin to feel a deep restlessness. The comforts of the hearth are no longer enough to hold me as I become consumed by the urge to travel and learn and draw another map.

Why do I make these maps? Because they give me pleasure and bring me money. True. But also because they seem to mirror my inward journeys of discovery and at the same time reflect my understanding of – and sense of place in – the outer world. At least that is the way I have come to see it. And I cannot yet envisage a time when the desire, or perhaps it is a need, will cease to be.

How far down the years did Roland's vision extend, I sometimes wonder? Perhaps this will turn out to be the pattern of our lives always. I should like to think that we will have our own child one day. But perhaps the one we share is all there is meant to be. Perhaps, perhaps . . . We must continue to seek our pleasures in the present – and in my life with Ellen and the little boy, they are not hard to find.

But what of the past? What was it that set us on our respective courses, helped each of us to shape our independence and become what we have become? It would be easy to say the pestilence. But the pestilence, I believe, merely provided the crucible in which a host of existing ingredients were first brought to blend. Similar alloys might well have resulted from different vessels on different flames. This much of the alchemist's lore I now understand. What I find less easy to comprehend is this: why, if the pestilence was indeed God's vengeance, did it bring us together, not destroy us?

But then there is a great deal that I still do not understand, and our own private miracle is no more than one small part of it . . .

Understand it or not, I will always remember vividly the great tide of tenderness and gentleness that washed us along through the days and weeks following the fire. We steeped ourselves in each other's stories, mapped out our recent pasts, made love, walked and played with the child, talked, talked, made love again and talked more.

I came to know of Ellen's anguish in the weeks after she had left the monastery; to suffer with her as she struggled to exist on what was left of her uncle's money, stopping a few days in this place, a week in that, each time hoping she had found somewhere she could settle, straighten her feelings and wits; each time realising that movement and activity remained the better balm. She described the sense of being drawn eastwards all the while, not to the home she was incapable of revisiting, but to the flat, open lands, the nearby cliffs and sea, the rich scent of loam and the tang of salt which seemed so familiar and comforting. As she began to speak of the child growing within her, she placed my hand gently on her belly so that I might imagine its gradual swelling, the strange little ripplings and flutterings beneath the smooth, taut skin.

In all the time we were apart, she said, I was never out of her thoughts: the hope that her message might reach me at the monastery had sustained her through the pain of her solitary childbirth and the grinding winter weeks that followed, as she had trudged the frozen streets to find work, then sewed through the night while the baby slept. And when, finally, she had been engaged as wet-nurse to the lace-merchant's child, the kindly attentions of the merchant's younger brother had served only to intensify her yearning for me. But most cruel of all, she told me, laying her head on my shoulder, had been the moment she saw me in the market-place with Peg. When, a little later, she had found the necklace in the street, she could no longer bear to be alone and had made her way in great distress to the lodgings of a friend on the outskirts of town. It was there, in the small hours, that word of the fire had reached her.

From where we now live, high on a wooded hillside, we cannot even see a village, let alone a town. When we first came, there was an old tumbledown dwelling in this overgrown clearing. I used the stones to build our own cottage and cut back the weeds, bushes and saplings to restore the sense of tranquil, enfolded space which must have delighted the previous inhabitants as much as it does Ellen and me. A stream chuckles through the clearing and rushes away down the steep hill in front of the cottage through a wide gulley filled with big, mossy boulders, where the boy likes to play for hours on end. The gulley makes a gap in the trees through which we can look westwards into the deep valley beneath and on beyond to a folding landscape of forest and moorland and a distant line of high, slate-blue hills.

I like to be high up when I am making my maps. I can see the land spread out around me, as I so often imagine it when transcribing its image; and the clean, clear air seems to sharpen my faculties. In the cold season this is especially true – and it has been very cold of late. Until yesterday, the snow had been falling steadily for some time. The roof of the cottage and the branches of the trees around us are laden. But in the afternoon the clouds began to shift and by evening there was a crisp, star sprinkled sky through which the geese came silently flying; so still it was that, for a moment, I imagined I could hear the ruffling of their pinions as they sailed through the clear, chill air. The glorious freedom of their passage never fails to stir me deeply. Perhaps it is because they, of all creatures, have least need of maps.

I was awake early this morning and left our bed as the first light came filtering through the shutters. Ellen did not stir, lying there with her dark, glossy hair fanned out on the pillow, her eyelashes fluttering faintly and the gentle, olive bloom of her cheeks inviting me to run my chill fingers there for softness and warmth. Ellen – whose love of life is so great that nothing she has had to endure could diminish it; whose generosity of spirit is matched only by that of one other person I have known, the very person whom she, when she realised that he would not live very much longer – and knew it himself – sought impulsively to comfort in the best way she could.

And curled into her side, the boy – who is so like his father to look at, with his fine pale features and deep grey eyes, that sometimes, when he gives me a particular look, I find myself wondering whether he is not really Roland reborn. At other times, I am aware of him being such a happy consequence of Ellen's deed that I wonder whether it was not all preordained. And when I consider that he is blessed with his mother's good health and common sense as well as his father's enquiring and intuitive nature, I warm even more to the notion. He bears his father's name, as well as his likeness. And it is for him and for the delight he brings me that I hasten home from my journeys, as much as for Ellen.

For a long while I stood and looked at them both, then pulled on warm clothes and went outside to bring in logs. The sun was coming up and beginning to touch the snow-laden forest on the far side of the valley with the faintest flush of pink. Our little clearing, however, was still in shadow. Walking to the dwindling woodpile I caught a glimpse of a fox trotting jauntily across the track behind the cottage. His brush was out, his ears pricked and he had the air of someone mightily satisfied with the night's work.

Our logs are stacked against the easterly wall of the cottage, out of the way of the prevailing wind which so often scours this high land with rain. I gathered an armful and glanced upwards at the fine array of icicles adorning the eaves. My attention was caught by one, larger than the rest, which seemed to be attached not to the timbers themselves but to something lodged beneath them.

I laid down the logs, then reached up for the icicle and grasped it as near as I could to its root. With only the slightest pressure it detached itself cleanly from beneath the eaves. I turned it upside down to look at the base and saw nuggets of what looked like mud embedded in the ice. I rubbed them with my finger and they crumbled easily, discolouring the ice with a gritty brown powder. Mud it was.

Whatever the icicle had been attached to was in deep shadow. I fetched a stool, climbed onto it, and felt beneath the eaves. Shortly I came into contact with a rough, round object not much bigger than my outstretched hand. I knew at once what it was – a swallow's nest.

I stepped from the stool and sat down on it, reconstructing the sequence of events in my mind. Drips of melting snow must have seeped between the shingles and formed icicles directly beneath. Those that reached the nest had continued to cling to it until they reached its lowest point and only there did the icicle begin to form. Being a large nest, it almost completely spanned one shingle and so collected drips from both sides: the more it froze, however, the more friable the mud would have become and sooner or later, given enough weight, the icicle would simply have broken free.

This was, of course, no proof of what had happened on the rain-barrel above the cowshed, but I found myself with a sudden image of a brilliant summer sky in which swallows swooped and dived around the great wooden receptacle, darting beneath its supporting timbers and hurtling out again from the shadows. It may have been a true memory, it may merely have been my imagination. But it offered a plausible solution to one mystery, at any rate.

I walked back into the cottage, carrying the logs and feeling more than a little pleased with my discovery.

I know that my father, practical man that he was, would have appreciated it, too.